Eagan Press Handbook Series

Emulsifiers

Clyde E. Stauffer

St. Paul, Minnesota, USA

TP
455
.S83
1999

Cover: FP90 dropping point apparatus courtesy of Mettler-Toledo, Inc., Columbus, OH 43240; ice cream and cake, ©1997 Artville LLC; coffee, cream, and muffin courtesy of Virginia Dare; center photo by DigitalVision.

Library of Congress Catalog Card Number: 99-61286
International Standard Book Number: 1-891127-02-0

©1999 by the American Association of Cereal Chemists, Inc.

All rights reserved.
No part of this book may be reproduced in any form, including photocopy, microfilm, information storage and retrieval system, computer database or software, or by any means, including electronic or mechanical, without written permission from the publisher.

Reference in this publication to a trademark, proprietary product, or company name is intended for explicit description only and does not imply approval or recommendation of the product to the exclusion of others that may be suitable.

Printed in the United States of America on acid-free paper

American Association of Cereal Chemists
3340 Pilot Knob Road
St. Paul, Minnesota 55121-2097, USA

About the Eagan Press Handbook Series

The Eagan Press Handbook series was developed for food industry practitioners. It offers a practical approach to understanding the basics of food ingredients, applications, and processes—whether the reader is a research chemist wanting practical information compiled in a single source or a purchasing agent trying to understand product specifications. The handbook series is designed to reach a broad readership; the books are not limited to a single product category but rather serve professionals in all segments of the food processing industry and their allied suppliers.

In developing this series, Eagan Press recognized the need to fill the gap between the highly fragmented, theoretical, and often not readily available information in the scientific literature and the product-specific information available from suppliers. It enlisted experts in specific areas to contribute their expertise to the development and fruition of this series.

The content of the books has been prepared in a rigorous manner, including substantial peer review and editing, and is presented in a user friendly format with definitions of terms, examples, illustrations, and trouble-shooting tips. The result is a set of practical guides containing information useful to those involved in product development, production, testing, ingredient purchasing, engineering, and marketing aspects of the food industry.

Acknowledgments

Eagan Press thanks the following individuals for their contributions to the preparation of this book:

Mario Colombo, Uniqema, Brantford, Ontario, Canada

Timothy Cottrell, Uniqema, Brantford, Ontario, Canada

Andre J. Eydt, Lonza Inc., Annandale, NJ

Dilip K. Nakhasi, Stepan Company, Maywood, NJ

Larry Skogerson, American Ingredients Company, Kansas City, MO

Lawrence E. Werner, AC Humko Corp., Memphis, TN

Contents

1. Emulsions and Foams • 1

Surface Activity: surfactants as amphiphiles • surface free energy and interfacial tension • surface excess of emulsifier • measuring surface tension

Formation, Stabilization, and Wetting: formation of emulsions and foams • emulsion stabilization • foam drainage and film breakage • wetting of solid particles

Microemulsions

2. Molecular Organization • 15

Fat and Emulsifier Crystals: triglyceride crystals • emulsifier crystals • crystal modifiers

Mesophases: mesophase structures • surfactant phase diagrams

Significance for Food Applications

3. Food Emulsifiers • 25

Emulsifier Types: monoglycerides • monoglyceride derivatives • sorbitan derivatives • polyhydric emulsifiers • anionic emulsifiers • lecithin

Hydrophilic/Lipophilic Balance: basic principle of the concept • experimental determination of HLB

Proteins: foaming agents • emulsifying agents

Regulations

4. Bakery Products • 47

Antistaling Agents: starch gelatinization • starch retrogradation • bread staling • emulsifier-starch complexation

Dough Strengtheners

Aeration Agents

Troubleshooting

5. Dairy and Nondairy Products • 67

Milk

Butter and Margarine: butter • margarine

Whipped Cream and Nondairy Whipped Toppings: whipped cream • nondairy whipped toppings

Ice Cream

Coffee Whiteners

Troubleshooting

6. Dressings and Sauces • 77

Polysaccharides at Interfaces: gums • modified starch • cellulose derivatives
Salad Dressings: pourable salad dressings • spoonable salad dressings
Mayonnaise
Reduced-Fat Dressings and Sauces
Troubleshooting

7. Beverages • 89

Flavor Emulsions: oil phase • emulsion stabilizers • emulsion preparation
Stability: creaming • flocculation • coalescence
Microemulsions
Troubleshooting

Glossary • 95

Index • 101

CHAPTER 1

Emulsions and Foams

Interfaces are ubiquitous features of foods. This is true during preparation of a cake batter or margarine blend, for example, as well as in finished products. Three specific kinds of interfaces are of particular importance in foods: liquid-liquid, or emulsions; air-liquid, or foams; and solid-liquid, or dispersions.

Controlling the physical nature of the interfaces is often crucial to making a high-quality food product and is frequently achieved by including *emulsifiers* (also called surfactants) among the ingredients. Emulsifiers may be components of an ingredient (e.g., egg yolk) or additives (e.g., monoglyceride).

From a technical point of view, the terms "emulsifier" and "surfactant" are synonymous. In practice, however, an emulsifier is usually considered to be a food-related material, while a surfactant is generally connected with other processes. For example, surfactants are added to detergents for washing and are used to aid in flotation during ore separation. Since this book is food-related, other uses of surfactants will not be discussed.

Surface Activity

SURFACTANTS AS AMPHIPHILES

"Surfactant" is a coined word (from "surface-active agent") applied to molecules that migrate to interfaces between two physical phases and thus are more concentrated in the interfacial region than in the bulk solution phase. The key molecular characteristic of a surfactant is that it is *amphiphilic*. The *lipophilic* (or *hydrophobic*) part of the molecule prefers to be in a lipid (nonpolar) environment, and the *hydrophilic* part prefers to be in an aqueous (polar) environment. The word "prefers" actually means that the thermodynamic *free energy* of the system is at a minimum when the lipophilic part is in an oil (or air) phase and the hydrophilic part is in water. If a surfactant is dissolved in one phase of an ordinary mixture of oil and water, some portion of the surfactant will concentrate at the oil-water interface; and at equilibrium, the free energy of the interface, called interfacial or surface tension (γ), will be lower than if the surfactant were absent. Putting mechanical energy into the system (e.g., by mixing) in a way that subdivides one phase will increase the total amount of interfacial area and energy. The lower the amount of interfacial free energy per

In This Chapter:

Surface Activity
  Surfactants as Amphiphiles
  Surface Free Energy and Interfacial Tension
  Surface Excess of Emulsifier
  Measuring Surface Tension

Formation, Stabilization, and Wetting
  Formation of Emulsions and Foams
  Emulsion Stabilization
  Foam Drainage and Film Breakage
  Wetting of Solid Particles

Microemulsions

Interfaces—Boundaries between two phases. Various types of interfaces occur in foods: solid-liquid, gas-liquid, gas-solid, and liquid-liquid (two immiscible liquids).

Emulsifiers—Molecules that promote and/or stabilize emulsification, i.e., dispersion of one liquid in another (nonmiscible) liquid.

Amphiphilic—"Both loving." Pertains to molecules that possess both lipophilic ("fat-loving") and hydrophilic ("water-loving") regions.

Lipophilic—"Lipid loving." Pertains to the nonpolar parts of molecules that dissolve readily in a nonpolar medium such as vegetable oil. Generally synonymous with "hydrophobic."

Hydrophobic—"Water hating." Pertains to the (nonpolar) parts of molecules that do not readily enter a polar medium such as water.

Hydrophilic—"Water loving." Pertains to the (polar) parts of molecules that readily dissolve or disperse in a polar medium such as water.

Free energy—The thermodynamic energy of a closed system. Absolute free energy is not easily measured, but the change in free energy when the system is changed (e.g., when oil is dispersed in water) is more easily established.

Discontinuous phase—The dispersed (internal) phase in an emulsion. In an oil-in-water emulsion, oil is the discontinuous phase.

Continuous phase—The undispersed phase of an emulsion. In an oil-in-water emulsion, water is the continuous phase.

Ionic strength—Measure of the ionic character of an aqueous solution of salts. Ionic strength plays a role in numerous physical phenomena such as conductance of electrical current, folding of protein molecules, and degree of repulsion of charged surfaces in water.

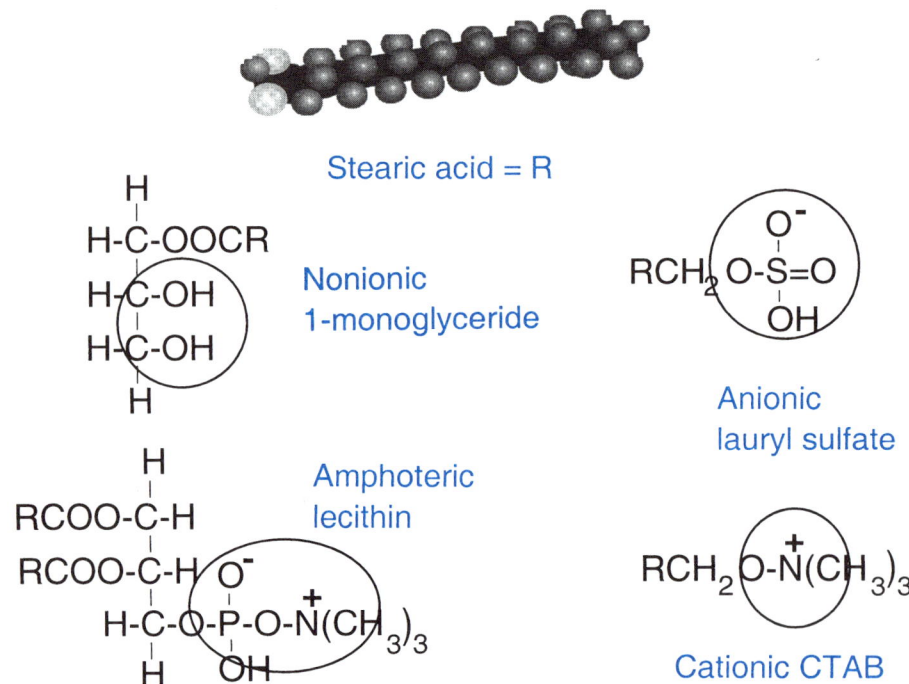

Fig. 1-1. Types of surfactants. The R group is a lipophilic hydrocarbon typified by stearic acid. CTAB = cetyltrimethylammonium bromide.

unit area, the larger the amount of new interfacial area that can be created for a given amount of energy input. The subdivided phase is called the *discontinuous phase* and the other the *continuous phase*.

As shown in Figure 1-1, surfactants have a lipophilic (fat-loving) and a hydrophilic (water-loving) part and thus are sometimes called amphiphilic (both-loving) compounds. The lipophilic part of emulsifiers (food surfactants) is usually a long-chain fatty acid obtained from a food-grade fat or oil. The hydrophilic portion is either nonionic (e.g., glycerol), anionic (negatively charged; e.g., lactate), or amphoteric (carrying both positive and negative charges; e.g., phosphatidylcholine). Cationic (positively charged) surfactants such as cetyltrimethylammonium bromide (CTAB) are usually bactericidal and somewhat toxic and are not used as food additives. Food surfactants include monoglyceride (nonionic), lauryl sulfate (anionic), and phosphatidylcholine or lecithin (amphoteric). The nonionic surfactants are relatively insensitive to pH and salt concentration in the aqueous phase, while the functionality of the ionic types may be markedly influenced by pH and *ionic strength*.

SURFACE FREE ENERGY AND INTERFACIAL TENSION

Increasing the amount of interface in a system requires work (energy) input. In a practical sense, this energy is supplied by shaking or mixing. If a stoppered bottle containing oil and water is gently inverted, only a small additional amount of interface (a few large oil drops) is generated. More vigorous shaking subdivides the oil droplets further; that is, it creates more interface. The system has a higher energy content; or in physical chemical terms, it has a higher total free energy. This concept is quantitated in the following manner. Consider a soap film contained in a rectangular wire frame (Fig. 1-2); the surface area of the film, A, equals $l \times x$. The right side of this frame, of length l, is moveable. The work necessary to move this side to the right by a distance dx is determined by the equation

$$\text{work} = 2\gamma l dx = 2\gamma dA \quad (1)$$

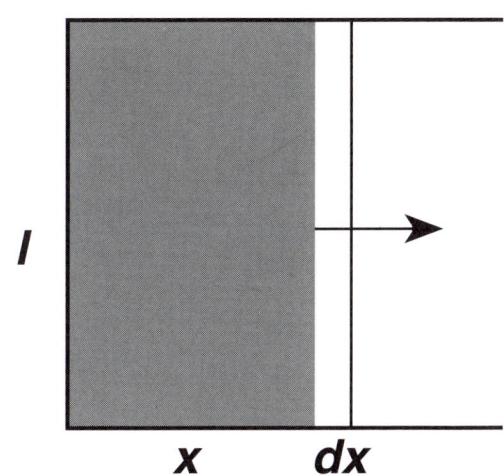

Fig. 1-2. Wire frame with one movable side (of length l) and containing an interior film. dx = distance moved.

in which γ is the free energy of the water-air interface (the factor 2 is used because both surfaces of the film must be considered). The total surface free energy equals $2\gamma l x$. The usual unit for γ is milli-Newtons per meter (mN/m), which is numerically equal to the older (non-SI) unit of dyn/cm or erg/cm^2. The concept is general; the surface could be the interface between two condensed phases, e.g., water and oil, in which case γ is interfacial tension. The terms surface free energy and *surface tension* are synonymous as are interfacial free energy and interfacial tension.

If the surface is curved (e.g., that of a droplet), the radius of curvature plays a role. Given an air bubble of radius r, the total surface energy is $4\pi r^2 \gamma$. Decreasing the radius by the amount dr decreases total surface energy by $8\pi r\gamma dr = 4\pi r^2 \gamma dr$. This change must be balanced by a pressure increase (ΔP), or the bubble would be compressed to nothingness. This pressure difference times the change in surface area equals the change in total surface energy:

$$\Delta P 4\pi r^2 dr = 8\pi r\gamma dr \quad (2)$$

and

$$\Delta P = 2\gamma/r \quad (3)$$

As indicated by equation 3, the internal pressure of a small bubble is greater than that of a large bubble. This has practical consequences in aerated food systems. In a cake batter, for example, time-lapse photography shows that small bubbles (containing carbon dioxide) disappear and large bubbles increase in size. The carbon dioxide in the small bubbles dissolves into the aqueous phase because of the higher inter-

Surface tension—The component of total free energy in a closed system caused by the presence of an interface.

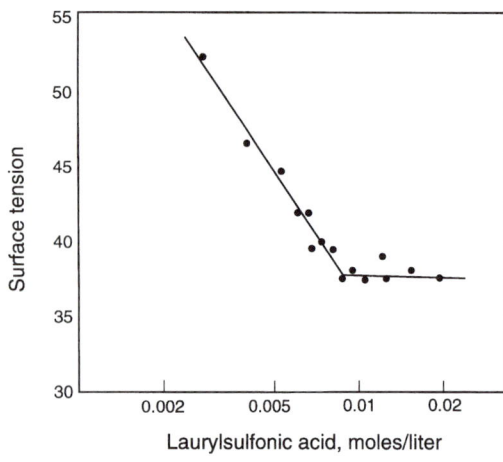

Fig. 1-3. Surface tension of laurylsulfonic acid solutions.

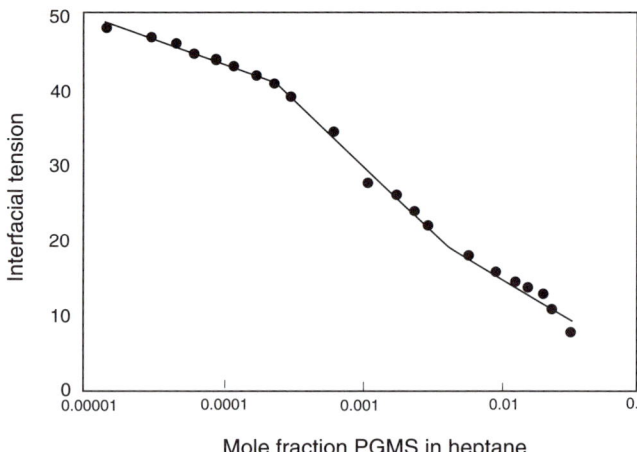

Fig. 1-4. Tension at the interfaces of water and solutions of propylene glycol monostearate (PGMS) in heptane.

Critical micelle concentration (CMC)—Concentration of a surfactant in aqueous solution at which colligative properties cease to change with increase in concentration.

Thermodynamic activity—A factor accounting for the fact that the concentration-dependent properties of dissolved molecules often deviate from a strictly linear dependence. For surfactants below the critical micelle concentration, this deviation is usually negligible.

Surface excess—The concentration of a surfactant in the interfacial region compared with its concentration in the bulk phase in which it is dissolved.

nal pressure and then enters the large bubbles, which are regions of lower internal pressure. Similar effects are expected in other foods in which the continuous phase may act as a conduit for dissolved gases.

The surface tension of a solution of a surfactant is lower than that of the pure solvent. Surface tension is roughly a linear function of ln(surfactant concentration) up to the *critical micelle concentration* (CMC) (Fig. 1-3). Above this concentration, the *thermodynamic activity* of the surfactant does not increase with the addition of more surfactant, and the surface tension remains constant. Interfacial tension also decreases with the concentration of an emulsifier dissolved in one of the phases. As shown in Figure 1-4, the decrease in γ, or interfacial tension, does not level off, because the emulsifier (propylene glycol monostearate) does not form micelles in the organic solvent phase (heptane). The changes in the slope of the plot are attributed to changes in the orientation of emulsifier molecules at the interface.

SURFACE EXCESS OF EMULSIFIER

Surfactant molecules concentrate at the interface; the lipophilic portion is in the nonpolar phase (air, organic solvent), and the hydrophilic portion is in the polar (water) phase. This migration of surfactant lowers the free energy of the total system, resulting in a higher concentration of surfactant in the region that includes the interface (Fig. 1-5). The difference between this concentration and the bulk concentration is called the *surface excess*, Γ.

The surface excess is calculated from a plot of interfacial tension (γ) versus the thermodynamic activity (a) of the surfactant:

$$\Gamma = -(a/RT)(d\gamma/da) \qquad (4)$$

in which R the universal gas constant and T is the temperature (°K).

For dilute solutions, surfactant thermodynamic activity equals concentration, and γ is calculated from the plot of γ versus molar concentration.

MEASURING SURFACE TENSION

Three things can be learned by measuring surface tension: 1) the minimum level of surface tension achievable; 2) the corresponding minimum CMC; and 3) the area covered by each surface-active molecule (from the slope of γ versus concentration curve) (Fig. 1-3).

Numerous ingenious methods have been devised to measure surface and interfacial tension. Space does not allow a full discussion of each method here, but selected methods will be described briefly. For more details and some of the tables of correction factors needed, see Chapter 1 of *Physical Chemistry of Surfaces* (1).

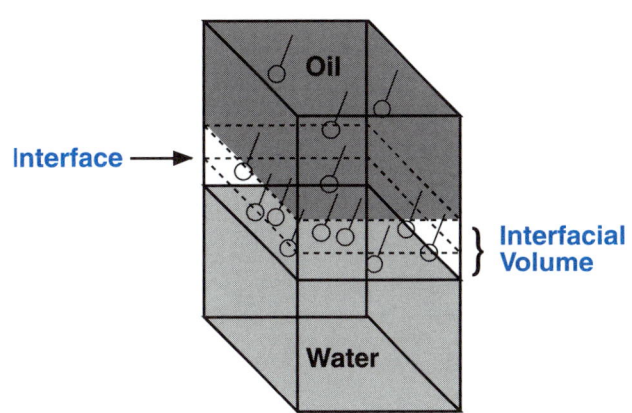

Fig. 1-5. Excess concentration of an oil-soluble emulsifier at the water-oil interface.

Capillary rise. If a clean glass capillary tube is held vertically and lowered to meet the surface of water or an aqueous solution, the liquid will rise in the tube because of the concave meniscus formed when the liquid wets the side of the tube. This curved meniscus results in a negative pressure (equation 3). The liquid rises until the weight of the column balances the negative pressure force. If the radius of the tube is known, the surface tension of the liquid can be calculated from the measured height of the column.

Drop weight. An ordinary laboratory burette with a tip of known diameter can be used. Liquid from the burette is allowed to slowly form a drop on the end of the tip. As the drop gradually increases in weight, it reaches a point at which gravity overcomes the surface tension holding the droplet to the tip. A portion of the drop separates and is caught in a tared container and weighed. By using a table of correction factors, the diameter of the tip, and the density of the liquid, surface tension is calculated.

Pendant drop. At some point before the drop separates, it has a bulbous shape with a somewhat narrow neck just below the tip of the burette. Measurements of the diameter of this pendant drop taken with an instrument such as a horizontal microscope at particular points along the vertical axis and use of published correction factor tables allow calculation of surface tension. This method is particularly useful for measuring interfacial tension by placing a drop of water into the oil being studied.

Ring tensiometer. A ring of known circumference, constructed of wire of known diameter, is attached to a balance arm or similar instrument for measuring vertical force on the ring. It is lowered into the liquid and then gradually raised through the surface. (More conveniently, the beaker containing the liquid is supported on a laboratory jack, and the beaker is slowly lowered.) As the ring is raised above the surface of the liquid, it carries a cylindrical film of liquid with it, and the force on the ring increases. At some point, the film ruptures. From the maximum force measured and with the aid of a table of correction factors, the surface tension can be calculated.

This method is also convenient for measuring interfacial tensions. The ring is immersed in the water, the oil is carefully layered on top, and the ring is then raised through the interface.

Wilhelmy balance. The Wilhelmy balance method is similar to that of the tensiometer in arrangement and operation, except that the ring is replaced with a cleaned glass slide, which is wetted by the water. Again, the maximum force is measured as the slide is slowly raised relative to the water surface. Surface tension is directly proportional to the total circumference (i.e., two sides plus two edges) of the glass slide, and empirical correction factors are not needed.

More sophisticated methods are available for measuring dynamic surface tension. As a practical recommendation, if surface or interfacial tension measurements are contemplated, the ring tensiometer method is probably the easiest for obtaining reliable results with a minimum of investment and operational difficulty.

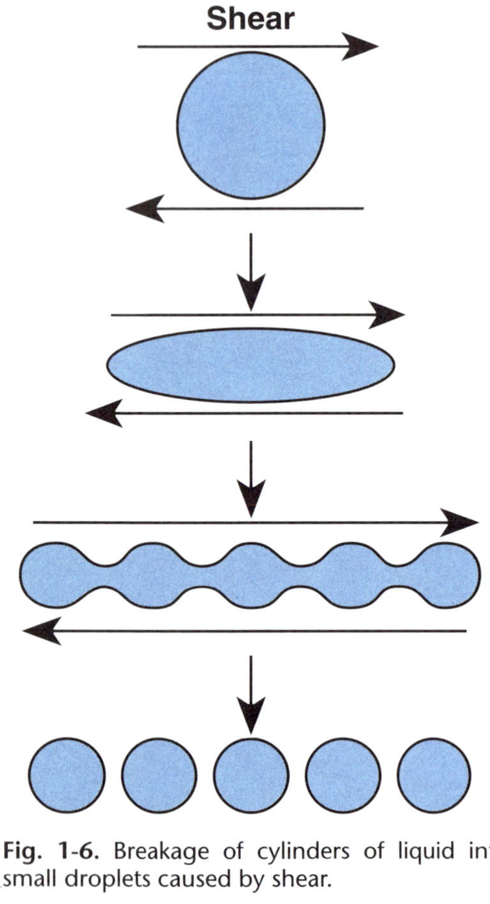

Formation, Stabilization, and Wetting

FORMATION OF EMULSIONS AND FOAMS

Simply adding oil to water does not result in an emulsion, because oil is a nonpolar liquid and water is polar. However, if the oil and water are shaken together, an emulsion forms. For example, in an oil/water emulsion, shaking causes the oil phase to separate into droplets that are dispersed in the water phase. The oil is called the dispersed phase, discontinuous phase, or internal phase, and the water is called the continuous phase.

Division of internal phase. Input of mechanical energy subdivides the droplets of the internal phase until a final average droplet diameter of 1–100 µm is reached. A cylinder of liquid whose length is more than 1.5 times its circumference is unstable and tends to break up into droplets. Mechanical stirring of an oil-water mixture forms drops that are then distorted into cylinders (along the lines of flow), which break up into smaller droplets (Fig. 1-6). The process is repeated until the droplets are

Fig. 1-6. Breakage of cylinders of liquid into small droplets caused by shear.

so small that they cannot be further distorted and further subdivision ceases.

A suspended liquid drop forms a sphere, because this shape has minimum surface area (hence, minimum interfacial free energy) for a given volume. Distortion is a flow shear effect, depending on droplet cross-section, which is related to the square of the radius. When droplet diameters are large, shear forces are greater than interfacial tension forces; droplets are distorted into cylinders; and subdivision occurs. Droplet radius decreases until the interfacial tension forces balance (or exceed) shear forces, and further division stops. In emulsification experiments in which the amount of mixing energy is constant and γ is changed by adding emulsifier, the average oil droplet diameter parallels γ; i.e., as more emulsifier is added, γ decreases and so does average droplet size. If γ is unchanged but mixing energy is increased, droplet size is also decreased. This is caused by the change in the balance of shear and interfacial forces, allowing cylindrical distortion of smaller droplets. Equipment design that enhances shear is more effective at dividing droplets.

Oil/water versus water/oil emulsions. If oil and water are mixed together and vigorously shaken, they form a dispersion of water droplets in oil and oil droplets in water. When shaking is stopped, the phases start to separate; small water drops fall toward the container bottom and oil drops rise. When like drops come into contact with each other, they *coalesce* and the emulsion quickly breaks down. Adding an emulsifier to the system changes the outcome. After standing, one phase becomes continuous, while the other (the discontinuous phase) remains dispersed. The nature of the emulsion is determined by the emulsifier. As a general rule, the continuous phase is the one in which the emulsifier is soluble. Thus, sodium stearate promotes an oil-in-water (O/W) emulsion in which the oil is dispersed in the continuous water phase, and zinc distearate promotes a water-in-oil (W/O) emulsion. Several qualitative theories have been advanced to explain this empirical rule.

> ### Box 1-1. Is It an O/W or a W/O Emulsion?
> During early testing of systems, the question sometimes arises about the nature of the emulsion formed. To easily determine the answer, fill a beaker half way with the emulsion. Add half that amount of water, stir gently, and let the beaker sit a few minutes. If it is an oil-in-water emulsion, the water will dilute the continuous phase and the contents of the beaker will appear uniform. If it is a water-in-oil emulsion, the added water will collect on the bottom and the emulsion will float on top of the water layer.

Coalesce—To combine; usually refers to two liquid (e.g., oil) drops combining into one drop.

The oriented wedge theory states that the emulsifier at the interface is wedge shaped (Fig. 1-7). The ionized end of a sodium soap has a wider (effective) radius than the hydrocarbon chain; hence, the oil-water interface should be curved, with the convex side toward the water phase. This favors formation of oil droplets and results in an O/W emulsion. However, the polar end of zinc distearate is smaller than the two hydrocarbon chains, the interface is convex toward the water phase, and a W/O emulsion is formed.

A second theory considers the relative ease with which the two types of droplets can coalesce. When a mixture is shaken, drops of both phases are formed. Sodium stearate ionizes, and the *electrical potential* hinders approach and coalescence of oil droplets. Water droplets, however, experience no such hindrance and readily touch and coalesce. Zinc distearate does not ionize and therefore does not interfere with the mutual approach of oil droplets, but *van der Waal's forces* favor water coalescence. Thus, the type of emulsion formed depends on the relative kinetics of oil-oil and water-water coalescence.

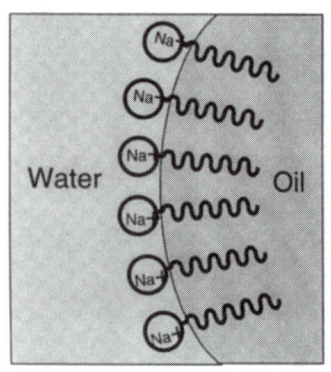
Oil-in-Water Emulsion

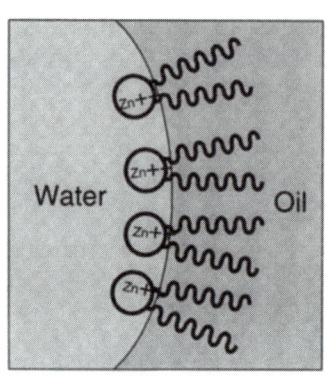
Water-in-Oil Emulsion

Fig. 1-7. Effect of wedge-shaped surfactant molecules on the convexity of the oil-water interface.

Foaming. Food foams are usually made by whipping an aqueous solution of a foaming agent such as a protein (e.g., egg white) or an emulsifier (e.g., one of the polysorbates). Air is first entrained by the action of the mechanical element (paddle, whip, or mixer blades), and then air bubbles are elongated and subdivided into smaller bubbles, just as described above for liquid internal phases.

Air is a nonpolar medium. Surfactants concentrate at the air-water interface, and the hydrophobic portion extends into the gas phase. In proteins (common foaming agents in foods), some amino acid side chains are hydrophilic while others are lipophilic. In their natural configurations, protein chains are usually folded so that the lipophilic residues are in the interior and the hydrophilic residues are on the surface. (A protein molecule has been described as "an oil droplet surrounded by a water-loving shell." This characterization has some merit as applied to relatively small albumin and globulin proteins.) When a protein such as egg albumen is exposed at an air-water interface, it tends to unfold, with the hydrophobic side chains entering the air phase and the hydrophilic chains remaining in the water phase (Fig. 1-8). If oil is also present, it will spread at the air-water

Electrical potential—The magnitude of electrical charge difference between two points.

van der Waal's forces—Short-range attractive forces between molecules resulting from the dipole moment of atoms in the molecules.

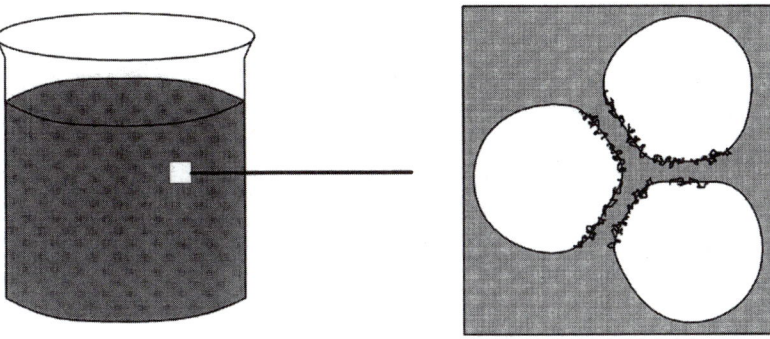

Fig. 1-8. Unfolding of protein at the air-water interface.

interface, displacing the protein and destabilizing the foam. Traces of oil (or egg yolk) in egg white, for instance, make it difficult if not impossible to whip the egg white into a foam. (For a more detailed discussion, see Chapters 3 and 5.)

EMULSION STABILIZATION

When two surfaces approach each other, two forces exist: one repulsive and one attractive. Whether or not the surfaces touch and coalesce depends on the relative strength of the two forces. This is equally true for liquids (e.g., oil droplets in an emulsion), solids (e.g., finely divided $CaCO_3$), and films (e.g., air bubbles in a foam).

Electrical repulsion forces. Electrical repulsion exists when the surfaces carry net charges of the same sign and the continuous phase is water. For example, if an O/W emulsion is stabilized by an anionic surfactant, the oil droplets have a negative charge on their surfaces. Electrical repulsion then tends to keep the droplets from making contact. At the oil surface, the electrical potential (or charge) is denoted by ψ_0. Cations are attracted into the region, partially neutralizing the surface negative charge. The value of ψ decreases as the distance from the oil drop surface increases and at some point becomes essentially zero. The rate of decrease of ψ is directly related to the ionic strength of the aqueous phase. Ionic strength (μ) is related to the concentration (c) of individual salt ions and the square of the ionic charge (z) of each ion:

$$\mu = \tfrac{1}{2} \Sigma c_i z_i^2 \tag{5}$$

Divalent ions are four times as effective as monovalent ions in decreasing ψ. Thus, $0.25M$ zinc sulfate, for example, is as effective as $1M$ sodium chloride in promoting emulsion *flocculation* or coalescence. If gravity is the only force bringing the droplets together, they will approach to a distance at which repulsion caused by ψ is just balanced by gravitational effects, and the emulsion will then be stable.

Flocculation—Collection of the internal phase of an emulsion or suspension; clumping. Flocculated materials are generally somewhat difficult to redisperse, unlike droplets in a creamed emulsion, which are easily dispersed by simple mixing.

Attractive forces. Attractive forces, collectively called van der Waal's forces, exist between two oil droplets. Simplistically, these forces may be thought of as the attraction between oil molecules at the O/W interfaces, which have lower energy when in contact with each other than when in contact with water. Several phenomena are involved. The most commonly considered, hydrophobic interactions and London dispersion forces, are unaffected by ionic strength. Suspensions of solids (e.g., cellulose fibers and finely divided $CaCO_3$) are stabilized in the same way. Ionic surfactants are used that selectively adsorb to the solid surface, generating a ψ potential and making possible a stable suspension.

Creaming. Creaming is the collection of the lighter droplets in the upper part of the mixture. Oil droplets in an O/W emulsion float to the top, because the density of vegetable oil is about 0.91 g/mL, 0.08 g/mL less than that of water. The rate at which they rise depends on particle diameter. A drop with a diameter of 1 µm rises at a rate of 4 cm per day, while one with a 10-µm diameter rises 4 m per day. Obviously, reducing the average droplet size reduces the rate of creaming. Fat globules in raw milk have diameters of 0.2–20 µm. After homogenization, the average diameter is <1 µm. In raw milk, the average flotation rate is 36 cm per day, while in homogenized milk it is 1 cm per day. Creaming brings the oil droplets closer together, and if contact is not prevented (e.g., by ionic repulsion), coalescence occurs. A simple creamed layer of stabilized oil droplets is readily redispersed by inverting the container a few times.

Ionic stabilization. As discussed above, two surfaces carrying like charges repel each other. The thickness of the electrical double layer (the region where $\psi > 0$) is affected by ionic strength. As long as ionic strength is low, electrical repulsion is greater than van der Waal's attraction, and the droplets remain suspended. If, by addition of salt (particularly divalent or trivalent salts), the ψ potential is markedly suppressed, the surfaces can approach to the point at which van der Waal's forces override repulsion, and the droplets can touch and coalesce. At some intermediate ionic strength, the two forces are approximately equal, and the droplets will remain separated by about one droplet diameter. A practical conclusion can be drawn from this: if the emulsifier used is ionic in nature, the salt concentration of the aqueous phase markedly affects emulsion stability. Low salt concentration enhances stability, while high salt concentration increases flocculation and/or coalescence.

Steric hindrance. Another form of stabilization is relatively independent of ionic strength: the oil droplets are prevented from making contact by simple steric hindrance. This may take one of two forms, either an immobilized water layer at the interface or a solid interfacial film. Emulsion stabilization by proteins, gums, and polyoxyethylene derivatives occurs by the first mechanism. Hydrophobic parts of the stabilizers adsorb at the oil surface, but adjacent large hydrophilic segments are hydrated and form an immobilized layer approximately of 10–100 nm thick (Fig. 1-9). As mentioned, these hydrated segments frequently interact to cause flocculation while coalescence of the oil drops is prevented. Such emulsions are frequently used as carriers for oil-soluble flavors, essences, and colorants.

The α-tending emulsifiers such as propylene glycol monostearate are oil soluble. The emulsifier adsorbs at the oil-water interface, but under certain conditions (such as low temperature or the presence of a free fatty acid), the emulsifier forms a solid interfacial film (Fig. 1-10). While the oil droplets may make contact, the film prevents coalescence. The interfacial layer actually appears to be crystalline and has a well-defined melting point.

Flocculation. In some emulsions, particularly those with high molecular weight emulsifying agents (e.g., proteins and gums), the oil drops rise to the top and form a layer that is rather resistant to redispersion. The size (and number) of the oil droplets does not change; coalescence does not occur. The material adsorbed at the interface interacts, probably through hydrogen bonds and perhaps some ionic bonds, to hold the droplets together. The phenomenon is termed "flocculation" or "clumping." One positive application of this process is in water treatment plants, where certain surface-active materials adsorb to solid impurities, promoting flocculation and allowing easy removal by filtration. Flocculation is much less significant in food applications.

FOAM DRAINAGE AND FILM BREAKAGE

Considering the forces involved, a foam is very similar to an O/W emulsion. The terminology is somewhat different, but the results are the same: either a foam is stable or the gas bubbles coalesce and the foam breaks down. Rather than creaming, a foam is said to "drain." The effect is the same: the water phase concentrates at the bottom of the container and the dispersed phase at the top. The *volume fraction* (ϕ) of gas in a foam is usually much higher than the ϕ of oil in an emulsion. Whipping egg white, for example, may easily give a 10- to 15-fold expansion (ϕ = 0.9–0.93), while mayonnaise (ϕ = 0.7–0.8) is the food emulsion with the highest oil content.

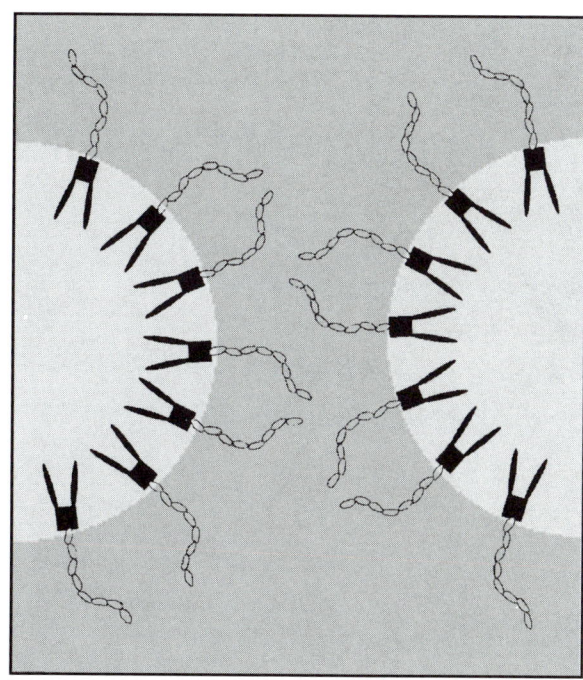

Fig. 1-9. Immobilization of water by hydrophilic chains at the oil-water interface.

Fig. 1-10. Interfacial film formed by an α-tending emulsifer. A water drop was suspended in vegetable oil containing 10% propylene glycol monoester. After a few minutes, some of the water was withdrawn.

Volume fraction—The fraction of total volume of a system represented by one of the discrete phases.

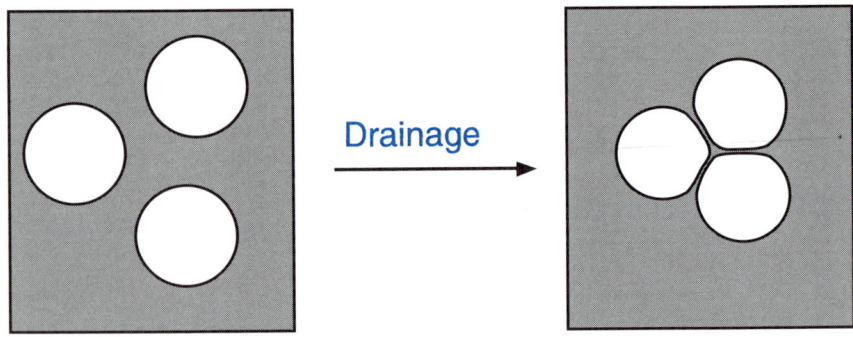

Fig. 1-11. Plateau's border formed at the juncture of air bubbles in a foam.

Plateau's border—The point at which three or four gas cells nearly touch in a foam.

Brownian motion—The random, thermal movement of minute, solute particles observable under a microscope and caused by collision of solvent molecules with the particles.

The mechanism of air incorporation and subdivision in a foam is the same as that of an emulsion; large bubbles are elongated, and the unstable cylinders spontaneously divide. In a wet foam ($\phi < 0.7$), the initial drainage of liquid from the regions between the bubbles is caused by gravity and is governed primarily by viscosity. Figure 1-11 shows how the film between three bubbles meets at a 120° angle after drainage. In three dimensions, four touching bubbles meet at the tetrahedral angle of 109°28'. In the real situation, the bubbles assume the shape of regular polyhedra, but the contact angles remain fairly close to the ideal value. The thicker liquid in the corners, known as *Plateau's border* (named for the Belgian physical chemist J. Plateau) has a pressure lower than that of the straight contact films, and liquid moves into these regions. Drainage of a dry foam ($\phi > 0.7$) is probably through these borders, connected throughout the foam.

Foam stability and emulsion stability are governed by similar factors. Thus, in a soap foam, the negative charges located at the air-water interface result in repulsion as the two surfaces of the film approach each other, and drainage stops when an equilibrium film thickness is reached. This thickness is influenced by the ionic strength of the aqueous phase; increasing the ionic strength gives foams of lower stability. Protein stabilizes foams by a combination of steric hindrance and surface viscosity. When egg whites are whipped, the protein molecules unfold, with the hydrophobic side chains entering the air phase and the hydrophilic chains remaining in the water phase. In Figure 1-8, the heavy lines in the magnified section represent unfolded albumin proteins adsorbed at the air-water interface. The portion of the proteins located in the aqueous phase hold water, preventing it from draining away from this region and hence preventing the air bubbles from coalescing and destabilizing the foam.

Film breakage is thought to result from random fluctuations (e.g., *Brownian motion*) that momentarily bring the two surfaces into contact, allowing the air bubbles to merge. These fluctuations are minimized when the surface viscosity is increased. The addition of an alcohol to a soap solution (e.g., dodecanol added to sodium laurate) increases surface viscosity and also increases foam stability. Surface viscosity of some (but not all) protein solutions is quite high, and there is some correlation between this property and the ability of the protein to give a stable foam. Bulk viscosity does not correlate with stability of drained films, but if a wet foam is desired, increasing bulk viscosity (e.g., by adding a high-viscosity gum) extends the usable life of the foam by slowing the drainage rate.

WETTING OF SOLID PARTICLES

Some surfactants are good *wetting agents*. This property is useful in many circumstances: enhancing dispersion of dry mixes in liquid, improving spreadability of chocolate and cocoa-based coatings, and incorporation of dietary fiber materials into dressings. Qualitatively, a drop of water is placed on the solid surface. If the contact angle $\theta < 90°$ (Fig. 1-12), the water does not spread; it is said that the solid is not wetted. If $\theta > 90°$, the water spreads, and the solid is wetted.

The angle θ is determined by the surface tension at the three interfaces involved:

$$\cos \theta = (\gamma_{SV} - \gamma_{SL})/\gamma_{LV} \qquad (6)$$

The spreading coefficient ($S_{L/S}$) is defined as

$$S_{L/S} = \gamma_{SV} - \gamma_{SL} - \gamma_{LV} \qquad (7)$$

in which S is solid, L is liquid, and V is vapor.

When $S_{L/S} > 0$, wetting occurs, and the liquid spreads. An efficient wetting agent is one that minimizes the surface tension of the air-water and solid-water interfaces while leaving the air-solid surface tension unchanged. This is the situation, for example, when dry beverage powder is added to water. Sodium lauryl sulfate lowers the air-water and solid-water interfacial tensions and enhances dispersibility. In the absence of the surfactant, the contact angle at many of the (irregular) solid surfaces is <90°, and the powder with its entrapped air floats on the top of the water.

> **Wetting agents**—Agents that promote spreading of a liquid on a solid surface.

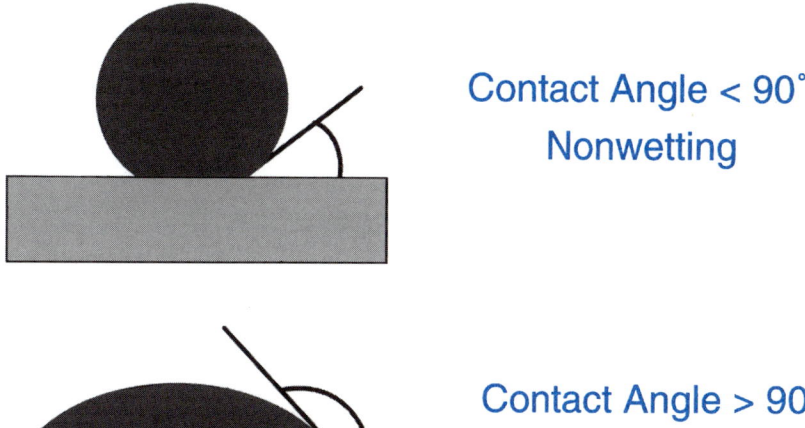

Contact Angle < 90°
Nonwetting

Contact Angle > 90°
Wetting

Fig. 1-12. Contact angle at the liquid-solid-air juncture of nonspreading and spreading liquid drops.

In chocolate coating, the liquid phase is an oil (cocoa butter). The addition of lecithin aids the wetting of solid cocoa particles by this oil, most probably by lowering γ_{SL}, thereby lowering the viscosity of the heterogeneous mass and giving a smoother mouthfeel to the final product.

Visible light—Light visible to the human eye. The wavelength of visible light is approximately 400–700 nm.

Microemulsions—Emulsions in which the diameters of the droplets in the dispersed phase are much smaller than the wavelength of visible light. The droplets do not scatter light; hence, the emulsion appears transparent.

Microemulsions

It is possible to make emulsions in which the diameters of the oil droplets range from 1.5 to 150 nm. The droplet size is less than the wavelength of *visible light*, and the emulsion appears transparent because no light scattering occurs. Several different strategies for making *microemulsions* have been tried, but a simple example in which mineral oil and water are used demonstrates the principles involved. With pure liquids, γ is 41 mN/m, but the inclusion of $0.001M$ oleic acid in the water phase reduces γ to 31 mN/m, and a moderately stable emulsion may be formed. Neutralization of the acid with $0.001M$ NaOH lowers γ to 7.2 mN/m and gives a stable emulsion. Making the water phase $0.001M$ in NaCl further lowers γ to less than 0.01 mN/m. This system will spontaneously form an emulsion; Brownian movement provides sufficient shear forces to elongate droplets into cylinders and cause further subdivision.

Spontaneous emulsions such as this one are frequently opalescent, because some particles have diameters approaching 400 nm, the wavelength of light. Transparent microemulsions generally require a surfactant plus a cosurfactant, for example, acetyl monoglyceride plus hexanol. For use in food, various polyglycerol esters have shown some promise for making W/O microemulsions. The technology is promising but needs further refinement before it can be readily applied to food systems.

Reference

1. Adamson, A. W. 1967. *Physical Chemistry of Surfaces*, 2nd ed. Interscience Publishers, New York.

CHAPTER 2

Molecular Organization

Fat and Emulsifier Crystals

Solid materials can be either *crystalline* or *amorphous*. In a crystalline solid (e.g., salt), the molecules are arranged in a repeating three-dimensional pattern. An amorphous material (frequently called a glass) has no such internal organization. A beam of *photons* (most often *X rays*) directed onto a crystalline material is refracted into a regular pattern that can be recorded (e.g., on a photographic film) and analyzed to obtain the dimensions of the crystal *unit cell*. From this information, the orientation of individual molecules is often inferred. If the object of X-ray analysis is a single crystal (not a powder), even more information can be obtained, and often the location of individual atoms within the unit cell can be determined.

TRIGLYCERIDE CRYSTALS

Solid *triglyceride* molecules resemble an elongated "h" (Fig. 2-1). The two parallel chains are the fatty acids at positions 1 and 3 of glyc-

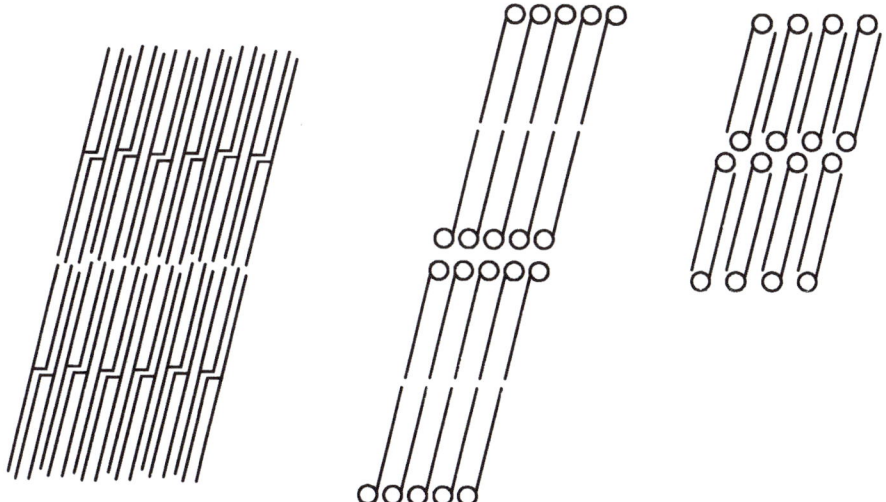

Fig. 2-1. Chain orientation in fat crystals. Left, paired, double chain length (DCL) orientation of a saturated triglyceride. Center, DCL orientation typically found in a saturated monoglyceride. Right, single chain length (SCL) orientation found in unsaturated monoglycerides and many emulsifiers that have relatively large hydrophilic groups.

In This Chapter:

Fat and Emulsifier Crystals
  Triglyceride Crystals
  Emulsifier Crystals
  Crystal Modifiers
Mesophases
  Mesophase Structures
  Surfactant Phase Diagrams
  Significance for Food Applications

Crystalline—Pertaining to a state in which atoms or molecules are arranged in an ordered three-dimensional array. Long-range order is discerned by X-ray analysis.

Amorphous—Pertaining to the random arrangement of atoms or molecules with no discernible long-range order.

Photon—Basic unit of light waves.

X rays—A region of light waves with short wave length and high-energy photons.

Unit cell—The smallest ordered unit of a crystal. It may contain only a few atoms (e.g., sodium chloride) or several molecules (e.g., a fat unit crystal).

Triglyceride—Lipid with three fatty acids esterified to a glycerol molecule.

erol; the single chain length (SCL), of course, is the chain attached at position 2. This is sometimes referred to as the "tuning fork" orientation. In the crystal, the molecules are paired as shown, giving a unit cell repeating dimension equal to two fatty acid chains, the double chain length (DCL) configuration. A triple chain length (TCL) configuration is possible in which the single chains of the tuning fork are paired, but this is rarely seen in natural fats. When it does occur, it is usually when the fatty acids vary by four or more carbons in chain length.

When a melted fat is rapidly cooled, it solidifies into a waxy material (resembling paraffin wax) termed "α crystals." If cooling is extremely slow, the triglycerides in the fat with the highest melting point have time to form stable β crystals. With intermediate cooling rates, the fat first forms α crystals, which rather quickly melt and reform into the *metastable* β′ crystals. The difference between the three crystal types has to do with the arrangement (crystal packing) of the pairs of fatty acid chains. The order of melting points of the crystal forms is α < β′ < β. For pure tristearin (glycerol tristearate), the melting points are 54.7, 63.2, and 73.5°C, respectively. Other factors that govern fat crystal stability and functionality have been discussed at greater length (1).

Diglycerides exist in two isomeric forms: the 1,2-diglyceride and the 1,3-diglyceride. The crystal structures of these materials have not been studied extensively, but it appears that the 1,2 isomer is stable in the β′ form, while the 1,3 isomer is stable as a β crystal. It has been suggested that 1,2-diglycerides can be used to stabilize shortenings and margarines in the desired β′ crystal structure, which imparts smoothness and plasticity to the fat. Unfortunately, the 1,3 isomer is thermodynamically more stable than the 1,2 isomer, and isomerization occurs at moderately elevated temperatures. No commercial form of 1,2-diglyceride is available today.

EMULSIFIER CRYSTALS

The structure of triglyceride crystals is governed by the nature of the fatty acids (saturated or unsaturated and varying chain length) and how these chains interact (fit together) in three dimensions. In emulsifier crystal structure, however, the predominant factor is the hydrophilic moiety, which is relatively a much larger part of the molecule. The size of the hydrophilic group, as well as the extent and spatial distribution of hydrogen bonding between adjacent groups, has a much greater influence on molecular packing in the crystal than does the nature of the fatty acid chain. A simple emulsifier such as a *monoglyceride* (e.g., glycerol monostearate [GMS]) generally crystallizes in the double chain configuration (Fig. 2-1), while those with larger hydrophilic groups (e.g., lactylated monoglyceride) more often crystallize in the SCL configuration.

1-Monoglycerides exhibit *polymorphic behavior* similar to that of triglycerides. When cooled from a melt, they first form α crystals, and

---

Metastable—Pertaining to a physical state that is not at the lowest possible free energy for the system. The activation energy for transition to the more stable state is high enough that the system remains in the current state for a significant length of time.

Diglyceride—Lipid with two fatty acids esterified to a glycerol molecule.

Monoglyceride—Lipid with one fatty acid esterified to a glycerol molecule.

Polymorphic behavior—Ability of a material to crystallize in more than one three-dimensional arrangement.

then on holding at intermediate temperatures, they transform to the β crystal form. Many emulsifiers (called α-*tending emulsifiers*) are stable in the α crystalline form; special techniques (such as crystallization from a solution) must be used to get the stable β form of these materials. Monoglycerides are usually found in the DCL configuration, while numerous other emulsifiers are more stable in the SCL configuration. One cannot generalize about the effects these various crystal tendencies have on functionality in food systems. Situations in which a specific crystal form plays a role are discussed later.

Finally, it should be noted that hydrogen bonding between hydrophilic head groups imparts stability to emulsifier crystals. The melting point of a saturated monoglyceride is typically 5–10 degrees C greater than that of the corresponding triglyceride. Even when monoglycerides are melted, X-ray diffraction shows the presence of a certain amount of structure corresponding to the length of the SCL configuration.

> α-**Tending emulsifier**— Emulsifier that forms a solid film at the oil-water interface under proper conditions of temperature (low) and concentration (high).

CRYSTAL MODIFIERS

Two types of modification of crystal behavior of fats are of interest to food technologists: 1) inhibition of crystal formation to prevent "clouding" (crystallization of solid fat) in salad oils; and 2) inhibition of crystal polymorphic changes, e.g., β′ to β crystal transformation in shortening and margarine. Emulsifiers are effective in both of these roles, although different ones are used for each of the functions.

When salad oils are cooled (held in the refrigerator), the saturated fat triglycerides begin slowly to form crystals. If the product is a simple oil and vinegar dressing, this is merely unsightly, but if the food is an emulsion such as mayonnaise, the crystals can actually lead to breakdown and separation of the oil phase. To prevent this, the oil is often "winterized"; i.e., the solid fat triglycerides (generally no more than 1–2% of the total oil) are extracted by chilling the oil and removing any crystals formed. However, removal is not usually complete, and a crystal inhibitor is also added to the oil. Commonly used crystal inhibitors are oxystearin and sorbitan monostearate. Other emulsifiers, for example, polyglycerol esters, sucrose esters, and polyoxyethylene sorbitan esters, are also effective in this role. Fat crystals grow by the addition of molecules (e.g., tristearin) to the surface of nuclei (small fat crystals) and then incorporation into the growing face of the crystal. Inhibitors also adsorb to the growing face, but since they do not match the dimensions of the triglyceride exactly, they interrupt the addition of more triglyceride molecules, keeping the crystal small (and essentially invisible to the eye). When the oil is warmed by removal from the refrigerator, many of the smallest crystals redissolve. Thus, in effect, the shelf life of the oil-containing food is extended.

Shortening and margarine are manufactured in the metastable β′ crystal form and have a smooth, plastic texture. If this crystal form is allowed to transform to the stable β crystal (e.g., by storage at somewhat elevated temperatures or by long-term storage), the product acquires a grainy, nonplastic texture that is considered unsuitable for most uses. Inhibition of the transformation (or to put it another way, stabilization in the β′ form) is a topic of interest to both manufacturers and users of these products. The fatty acid chains of the β crystal are somewhat more closely packed than those of the β′ crystal. If a small amount (1–3%) of an emulsifier (e.g., sorbitan monostearate or triglycerol-1-monostearate) is present, it cocrystallizes with the fat, but it does not fit precisely into the chain packing of the triglyceride. In other words, the emulsifier introduces a certain (minimal) amount of distortion into the crystal. Polymorphic transformation of triglycerides can be envisioned as a melting (of β′ crystals) and recrystallization (of β crystals), and the emulsifier interferes with this process. The end result is that the β′ crystals are stabilized and the shelf life of the product is extended.

Sorbitan tristearate is even more effective as a crystal stabilizer and is sold primarily for use in confectionery coatings. The development of *bloom* on chocolate-coated products is caused by a polymorphic transformation of the cocoa butter solid phase. Inclusion of an appropriate crystal modifier helps to counteract this unwanted change.

1-Monoglycerides are used in emulsified shortenings and margarines. However, 1-monoglycerides tend to hasten the β′ to β transformation, and the levels of other β′ stabilizers, such as sorbitan tristearate, must be increased to counteract this tendency.

Mesophases

Mixtures of surfactant and water form a number of different physical structures, depending on the surfactant-water ratio and the temperature. These mixtures are opalescent dispersions often called "liquid crystals," but they are more properly termed *mesophases*. This term (meaning "in-between phases") reflects the nature of the mixture. On a micro (molecular) scale, the surfactant and the water are separate phases, but on the macro scale (e.g., >1 μm), the mixture appears uniform and is stable (i.e., the phases do not separate).

MESOPHASE STRUCTURES

Micelles. Many emulsifiers are soluble in water to a significant degree. They exhibit all the colligative properties of dissolved materials, e.g., freezing point depression, boiling point elevation, and ability to conduct an electric current (if ionic); and, of course, they lower the interfacial tension of the water. At some concentration, however, these properties cease to change as more of the material is dissolved (Fig. 1-3.) This point is known as the critical micelle concentration (CMC) and is an important functional property of water-soluble surfactants.

A micelle is an aggregation of the emulsifier molecules, oriented with the hydrophobic chains to the inside and the hydrophilic

Bloom—A dusty, whitish appearance of the surface of chocolate coatings, caused by transformation of the fat crystals.

Mesophase—Opalescent or transparent liquid formed by a mixture of surfactant and water.

groups on the surface (Fig. 2-2). It may be thought of as an inverted *cubic mesophase*, with a high proportion of the continuous phase (water) present. Micellization is driven by free energy considerations and can be expressed as a reaction of emulsifier molecules Em:

$$x\ Em \leftrightarrow (Em)_x$$

The value of x varies with the type of emulsifier considered and may range from less than 100 to more than 3,000.

The study of micellization has contributed greatly to our understanding of hydrophobic interactions (also sometimes called hydrophobic bonds), an important concept in the realm of protein conformation, cell membrane structure, and other biochemical interactions. This topic is discussed with clarity by Tanford (2).

Lamellar mesophase. A monoglyceride such as GMS crystallizes in DCL bilayers. The thickness of each bilayer is defined by the length of two monoglyceride molecules end to end, which is about 5.5 nm in GMS (Fig. 2-3, left).

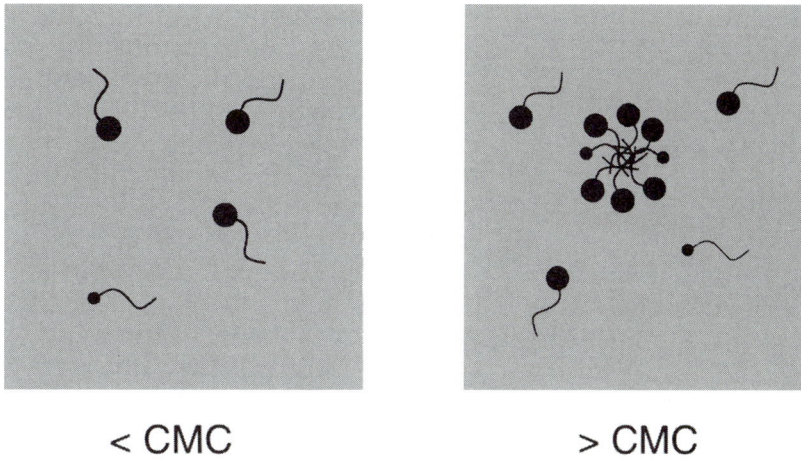

Fig. 2-2. Behavior of water-soluble surfactants below (<CMC) and above (>CMC) the critical micelle concentration (CMC).

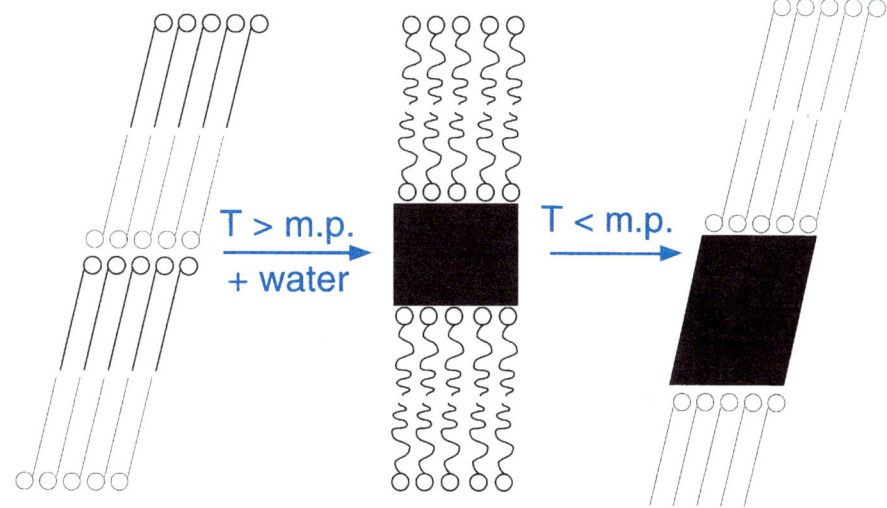

Fig. 2-3. Steps involved in the formation of lamellar mesophase with a saturated triglyceride. When the pure (distilled) monoglyceride is heated in the presence of water (left), it melts (fatty acid chains gain mobility), and water intrudes between the bilayer leaflets (center). Upon cooling, the chains solidify in the α-crystalline configuration (relatively unorganized), resulting in the α gel lamellar mesophase (right). T = temperature, and m.p. = melting point.

When mixed with water and heated, the crystals melt, the thickness of the bilayer decreases to about 3.8 nm (with concomitant lateral expansion), and water begins to intrude between the bilayers along the plane defined by the glycerol head groups (Fig. 2-3, center). This intrusion results in the formation of the *lamellar mesophase*. This material is rather fluid, but when the mixture is cooled, the lipid layers solidify in the α-crystalline state, and the material becomes a gel with a

Cubic mesophase—Mesophase in which spheres of water are found in a cubical arrangement in a matrix of the surfactant.

Lamellar mesophase—Mesophase characterized by bilayer leaflets of surfactant separated by layers of water.

lipid bilayer about 5.5 nm thick (Fig. 2-3, right). This phase is of particular interest to bakers, because there is evidence that the lamellar mesophase is the most efficient in promoting the interaction between monoglyceride and starch, producing the antistaling effects. Upon further cooling, the α-crystalline layers of monoglyceride transform into the more stable β-crystalline form. The van der Waal's attraction between the bilayers overcomes the tendency of the water to hydrate the glycerol head groups, and the water is expelled, yielding a suspension of β crystals in water.

With a simple nonionic saturated monoglyceride (i.e., GMS), the maximum thickness of the water layer in the mesophase is 1.6 nm, corresponding to 30% water. At this point, the osmotic pressure (favoring hydration of the head groups) is balanced by the van der Waal's attraction between the lipid bilayers. If more water is introduced, a dispersion of lamellar mesophase fragments in water is formed. Commercial distilled monoglycerides contain about 1% free fatty acid, and if this is neutralized with a base, the situation changes. The lipid-water interface takes on a negative charge, and electrostatic repulsion inhibits or slows the collapse and expulsion of water at the lower temperatures described above. The charged interface also favors the intrusion of more water into the space between the bilayers, and the thickness of the water layer increases in direct proportion to the amount of water added. At 75% water (an amount commonly found in commercial hydrated monoglyceride), the water layer is about 11 nm thick, twice the thickness of the lipid layer. As expected, this electrostatic stabilization can be counteracted by adding salts, and low concentrations (0.3%) in the aqueous phase will counteract the stabilizing effects of the anionic surfactant. Anionic monoglyceride derivatives, the succinate and diacetyltartrate esters, form lamellar mesophases under most conditions. This penchant is enhanced if the carboxyl group is partially neutralized so that the pH of a water dispersion of the surfactant is 4–6, the typical pH range for dough.

Cubic mesophase. At higher temperatures and water concentrations, the system may shift into the cubic mesophase structure (Fig. 2-4). The water is present as spheres totally surrounded by monoglyceride. This phase has a high viscosity and is sometimes called the *viscous isotropic phase* in the literature. In the presence of more water than can be accommodated in the internal spherical phase, a mixture of lumps of this cubic structure dispersed in excess water is obtained. With a saturated monoglyceride such as GMS, the lamellar structure is found most often under practical conditions; with unsaturated monoglycerides and lower temperatures, the cubic mesophase predominates. At lower water concentrations, the spherical water micelles are farther apart, so the viscosity of the mixture becomes lower, approaching that of melted pure surfactant. This is the fluid isotropic mesophase, sometimes referred to as the L2 phase.

Viscous isotropic phase—Another name for the cubic mesophase when the amount of water is sufficient to significantly raise the viscosity above that of the melted surfactant alone.

Hexagonal II mesophase. The third important structure, the *hexagonal II mesophase*, is formed by unsaturated monoglycerides and also by most of the other surfactants of interest to bakers. The hexagonal structure (Fig. 2-4) consists of rods of the internal phase arranged hexagonally within a matrix of the external phase. In the case of monoglycerides and sodium stearoyl lactylate, the internal phase is water, and the hexagonal II structure shown in Figure 2-4 is formed. With highly water-soluble, polyoxyethylene-type surfactants (e.g., polysorbates and ethoxylated monoglyceride), the internal phase is the lipophilic tail of the material and the external phase is water, resulting in the hexagonal I mesophase.

Liposomes. At higher water levels and within a limited temperature range, the lamellar mesophase is transformed into spherical multilamellar vesicles called *liposomes*. This process is sometimes called a lamellar dispersion and is most readily observed with phospholipids (lecithin). As water is added to the initial lamellar mesophase, the bilayers start to curve and ultimately bend and fuse, forming a hollow sphere. Water is trapped within the spheres and in the continuous phase surrounding them. Liposomes are most often used as delivery vehicles ("packaging") for drugs or cosmetic materials for topical application to the skin, where the phospholipid enhances penetration of the skin and delivery of the internal phase to the target site.

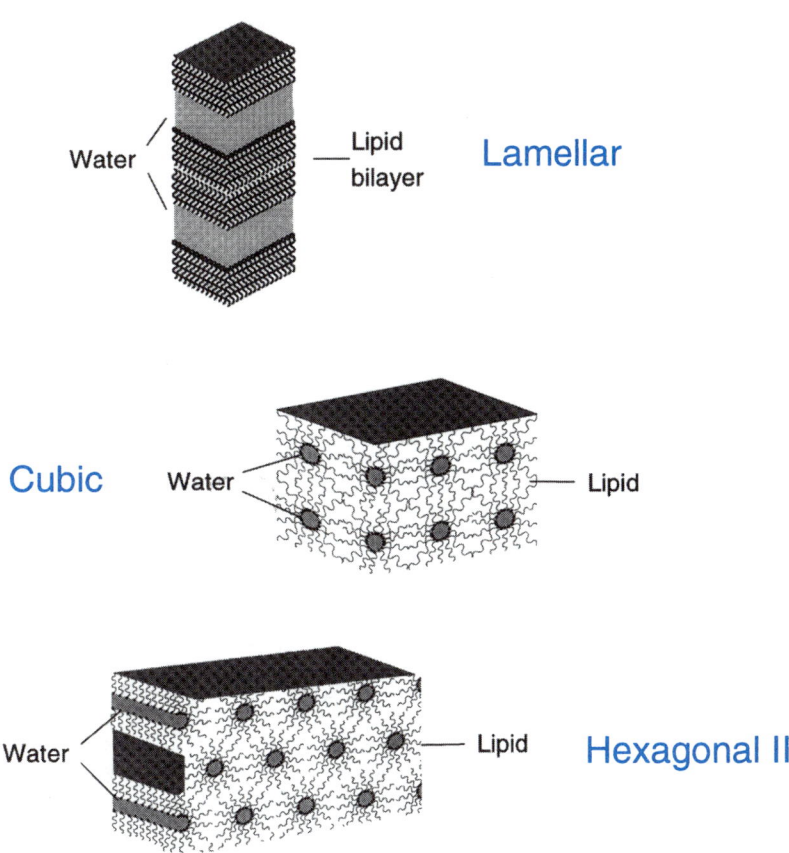

Fig. 2-4. Three types of mesophases formed by emulsifiers. In the lamellar mesophase, water is sandwiched between bilayers of emulsifier. In the cubic mesophase, spheres of water are regularly arranged within the emulsifier matrix. The hexagonal II mesophase consists of cylinders of water in the emulsifier matrix. The reverse of the cubic mesophase (spheres of emulsifier dispersed in a water matrix) is termed a "micellar dispersion." The reverse of the hexagonal array, termed "hexagonal I," has cylinders of emulsifier surrounded by water.

SURFACTANT PHASE DIAGRAMS

The exact mesophase structure of a particular surfactant in water has important implications in industries that use emulsifiers, and a great deal of research on the effect of water concentration and temperature has been done with commercial surfactants. The results are

Hexagonal II mesophase—Mesophase in which cylinders of water are arranged hexagonally in a matrix of the surfactant. When cylinders of surfactant are present with water as the matrix, it is called a hexagonal I mesophase.

Liposome—Hollow sphere formed by bilayers of surfactant, suspended in a water matrix, and containing water in the center.

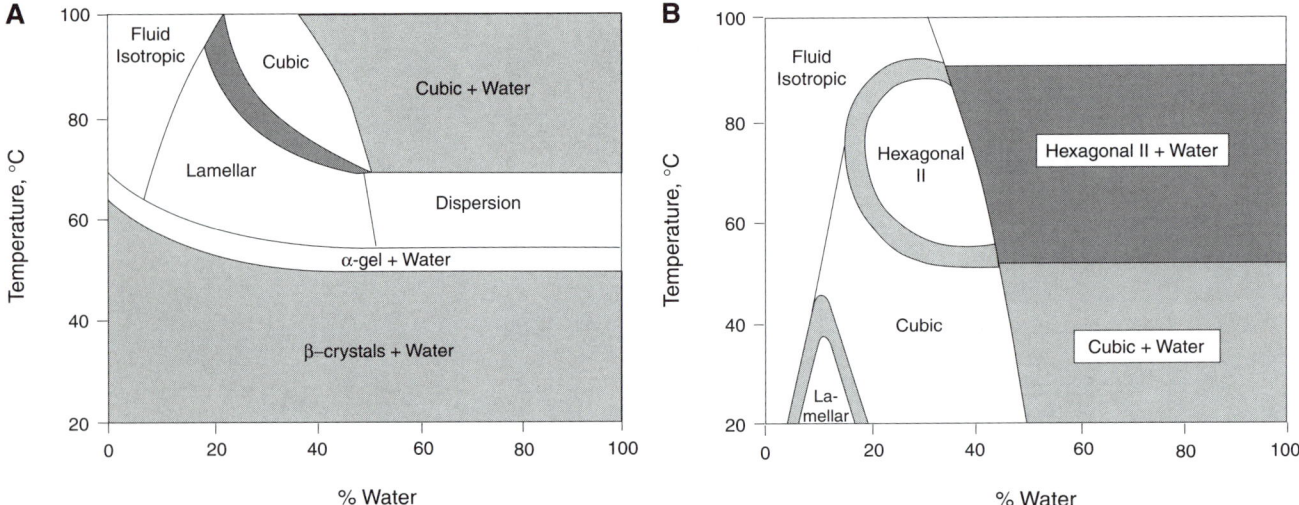

Fig. 2-5. Phase diagrams of two kinds of monoglycerides: saturated monoglyceride (distilled glycerol monostearate) (A) and distilled unsaturated monoglyceride (sunflower oil monoglyceride) (B).

usually shown as *phase diagrams,* with temperature on the *y*-axis and percent water on the *x*-axis. The interior of the graph is divided into regions that represent the various mesophases (Fig. 2-5). Phase diagrams provide guidance to researchers who are trying to produce surfactant systems that are functional under use conditions. They are used to characterize detergent systems and industrial applications of surfactants as well as food emulsifiers.

As shown in Figure 2-5A, a saturated monoglyceride forms a cubic mesophase at relatively low water concentrations (and above its melting point) and a lamellar mesophase at somewhat lower temperatures. Unsaturated monoglycerides, on the other hand, form predominantly the hexagonal II mesophase (Fig. 2-5B); the lamellar mesophase is found only at low water concentration and low temperatures. The inclusion of a nonpolar triglyceride (oil) in the system tremendously complicates the phase diagrams. For example, GMS forms a lamellar mesophase with water, as was noted above, but when soybean oil is added, the system converts into the hexagonal II structure.

Ternary phase diagrams from combinations of flour lipids, water, and salt have been studied. The structures listed above were all found in some region of the phase diagram. The studies are interesting, but as of this writing, the direct relevance to the practical application of surfactants in dough systems is not clear.

SIGNIFICANCE FOR FOOD APPLICATIONS

The mesophase structure of an emulsifier often influences its interaction with other food components such as starch and protein and hence its functionality in the food. The best-studied example of this is the interaction of monoglycerides with gelatinized starch. It is well known that the helical structure of starch can form a complex with

Phase diagram—A method of showing which mesophase structures are present at various water concentrations and temperatures.

linear hydrocarbons such as fatty acids. This is thought to be the basis for the antistaling activity of monoglycerides in bread. In numerous studies, Niels Krog (Grindsted Co., Denmark) has shown that this complexation takes place most readily when the monoglyceride is present in the lamellar mesophase. On the basis of this hypothesis, one can infer that the reason a saturated monoglyceride (such as GMS) is a more effective antistaling agent than an unsaturated monoglyceride is simply that there is a much larger range of temperature and water concentration where the lamellar phase is present. This can be seen by comparing the two phase diagrams in Figure 2-5.

Studies of the interaction of mesophases with proteins in foods are few, and most statements about such interactions are speculative rather than substantive. It has been found that wheat gliadin (a rather hydrophobic protein) is solubilized in the cubic phase of monoolein-water mixtures, but the significance of this for dough formation is unknown. Many of the emulsifiers used to strengthen dough, such as sodium stearoyl lactylate and polysorbates, form lamellar mesophases at the water content and temperatures found during dough mixing and proofing, and it has been suggested that in this form, they can contribute to the integrity (strength) of the gluten film. While lipid-protein interactions play a large role in the properties of many dairy foods, the significance of the mesophase behavior of the lipid has not been explored to any extent.

References

1. Stauffer, C. E. 1996. *Fats and Oils*. American Association of Cereal Chemists, St. Paul, MN.
2. Tanford, C. 1973. *The Hydrophobic Effect*. John Wiley & Sons, New York.

Supplemental Reading

Friberg, S. E., and Larsson, K., Eds. 1997. *Food Emulsions*, 3rd ed. Marcel Dekker, New York.

CHAPTER 3

Food Emulsifiers

Emulsifier Types

Actual commercial food emulsifiers are seldom exactly like the organic chemical structures that are discussed in this section. Rather, they are mixtures of similar compounds derived from natural raw materials. The hydrophobic fatty acid (or fatty alcohol) chain reflects the nature of the hydrogenated fat or oil used during manufacture. For example, glycerol monostearate (GMS) made from hydrogenated tallow has a saturated fatty acid composition of about 3% C14, 28% C16, 68% C18, and 1% C20, reflecting the chain length distribution in the source fat. If it is made from hydrogenated soybean oil, the chain length distribution is somewhat different (more C18 and less C16). In addition, the ratio of 1-monoglyceride to 2-monoglyceride varies depending on the temperature during manufacture.

Many emulsifiers are the result of rather complex condensation and polymerization reactions. Sorbitan monostearate is made by heating *sorbitol* and stearic acid together. Sorbitol cyclizes (dehydrates) to a mixture of *sorbitans* and *isosorbides*, which in turn is esterified to various extents by the stearic acid. By strict control of reaction conditions, the composition of the final product can be kept in a relatively narrow (and consistent) range, but it is still a mixture.

One should keep in mind, therefore, that the chemical structures shown here represent the major components in the commercial material and that related molecular species are also present.

MONOGLYCERIDES

Roughly 22 million kg (50 million lb) of monoglyceride are used annually in the United States in yeast-raised goods to retard *staling*. At least an equal amount finds its way into cakes, icings, and other applications. The third major use is in the manufacture of margarine. Overall, monoglycerides make up the single most important group for food uses, representing about 75% of total emulsifier production.

The use of monoglycerides in baking began during the 1930s, when "super-glycerinated shortening" became commercially available. Glycerin was added to ordinary shortening along with a small amount of alkaline catalyst. The mixture was heated, causing some interesterification of triglyceride with the glycerin, and the catalyst was removed by neutralization and washing with water. The resulting emulsified shortening contained about 3% monoglyceride and was

---

In This Chapter:

Emulsifier Types
  Monoglycerides
  Monoglyceride
    Derivatives
  Sorbitan Derivatives
  Polyhydric Emulsifiers
  Anionic Emulsifiers
  Lecithin

Hydrophilic/Lipophilic Balance
  Basic Principle of the
    Concept
  Experimental
    Determination of HLB

Proteins
  Foaming Agents
  Emulsifying Agents

Regulations

Sorbitol—Sugar alcohol produced by reduction of glucose, consisting of a chain of six carbons with a hydroxyl group on each carbon.

Sorbitan—Cyclic structure formed by linking the hydroxyls at the 1 and 4 positions of sorbitol through an ether linkage.

Isosorbide—Bicyclic structure formed from sorbitol, involving an ether linkage between hydroxyls at the 1 and 4 positions and hydroxyls at the 3 and 6 positions.

widely used for making cakes, particularly those containing high levels of sugar. The effectiveness of monoglyceride in retarding staling (crumb firming) in bread became known at about the same time, and bread bakers sought a more concentrated source of monoglyceride. This need was met by suppliers of plastic monoglyceride, which is made by increasing the ratio of glycerin to fat to achieve a final concentration of 40–60% monoglyceride. Most of the remainder is diglyceride. When industrial-scale molecular distillation processes were developed, it was logical to subject the plastic monoglyceride to this step, producing distilled monoglyceride containing a minimum of 90% (typically, about 95%) monoglyceride (the rest of the mixture is composed of diglyceride and small amounts of fatty acids and glycerol). The next stage was to make a lamellar mesophase product from this distilled monoglyceride, adding some anionic surfactant (usually sodium stearoyl lactylate [SSL]) to stabilize the hydrated monoglyceride, which contains roughly 25% monoglyceride, 3% SSL, and 72% water. More recently, manufacturers have developed a powdered distilled monoglyceride, in which the composition of the original feedstock fat is balanced between saturated and unsaturated fatty acids. The resulting powder is hydrated fairly rapidly during the process of dough mixing and is functional in complexing with gelatinized starch. Today, bakers use all three monoglyceride types (plastic, hydrated, and powdered distilled) with about equally good results.

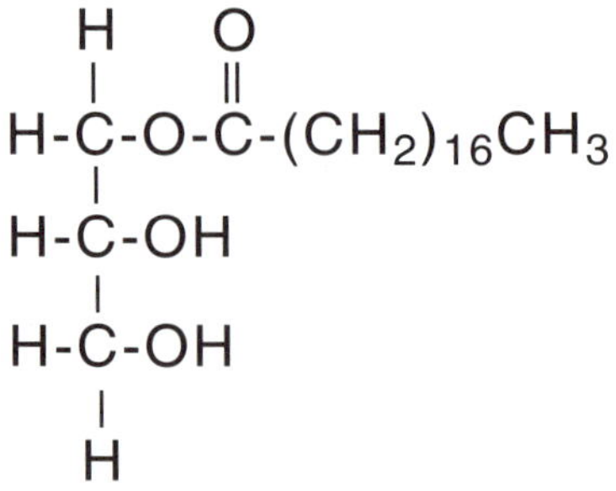

The monoglyceride structure shown in Figure 3-1 is 1-monostearin, also called α-monostearin. If the fatty acid is esterified at the middle hydroxyl, the compound is 2-monostearin, or β-monostearin. Both isomers are equally effective at retarding bread staling. In technical specifications, manufacturers usually give the monoglyceride content of their product as the percentage of α-*monoglyceride*. The routine analytical method for monoglyceride detects only the 1-isomer; quantitation of the 2-isomer requires the use of gas chromatography. The total monoglyceride content of a product is about 10% higher than the reported α-monoglyceride content. In a practical sense, however, when various products are compared for functionality and cost effectiveness, the α-monoglyceride content is a useful number, since for most products it equals about 91% of the total monoglyceride present.

The fatty acid composition of a monoglyceride reflects the makeup of the triglyceride fat from which it is made. Commercial GMS may contain as little as 65% stearate if it is made from fully hydrogenated lard or as much as 87% stearate if it is made from fully hydrogenated soybean oil. The other major saturated fatty acid is palmitic acid, and because complete hydrogenation (to an *iodine value* of 0) is not practical, a small percentage of unsaturated (oleic and/or elaidic) acid is also usually present. A typical commercial GMS has an iodine value of about 5. Iodine values for powdered distilled monoglycerides range from 19 to 36 and for plastic monoglycerides, typically from 65 to 75. The unsaturated fatty acids are a mixture of oleic and linoleic acids and their *trans* isomers. The phase diagram of a highly unsaturated

$$\text{H-C-O-C-(CH}_2)_{16}\text{CH}_3$$
$$|$$
$$\text{H-C-OH}$$
$$|$$
$$\text{H-C-OH}$$
$$|$$
$$\text{H}$$

1-Monostearin

Fig. 3-1. Glycerol 1-monostearate, also referred to as monostearin or GMS.

Staling—Phenomenon that occurs in baked products during storage. Stale product has a firmer crumb structure than fresh product; the crumb has a dry, harsh texture; and the flavor impact is significantly reduced.

α-Monoglyceride—Monoglyceride in which the fatty acid is esterified to the 1 position of glycerol. Esterification at the 2 position results in a β-monoglyceride.

Iodine value—A measurement of the number of double bonds in a fat or oil. A higher value means more double bonds.

monoglyceride is quite different from that of a saturated monoglyceride (Fig. 2-4), but the higher melting point of the partially hydrogenated monoglyceride (both saturated and unsaturated) makes its phase behavior much more like that of GMS than that of the sunflower monoglyceride. To the extent that mesophase behavior governs monoglyceride functionality, the plastic monoglycerides are quite adequate.

MONOGLYCERIDE DERIVATIVES

The derivatives of monoglyceride shown in Figures 3-2 and 3-3 are of two types: 1) *dough strengtheners* (succinyl monoglyceride [SMG], ethoxylated monoglyceride [EMG], and diacetyl tartrate ester of monoglyceride [DATEM]) and 2) α-*tending emulsifiers* (acetylated monoglyceride [AcMG], lactylated monoglyceride [LacMG], and

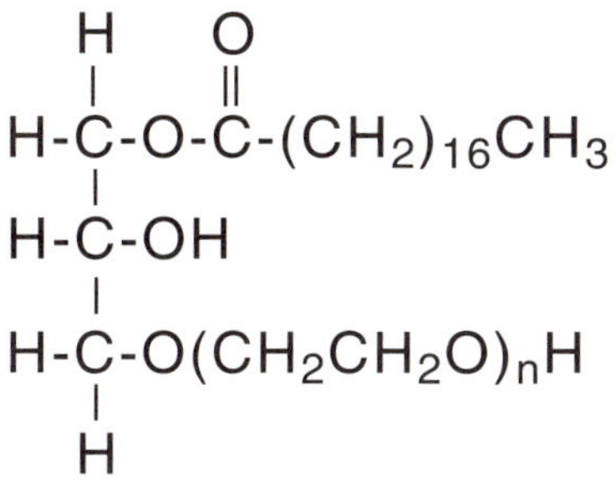

Dough strengthener—Material added to bread dough to increase the ability of the gluten to retain gas during proofing and baking.

α-Tending emulsifier—Emulsifier that forms a solid film at the oil-water interface under proper conditions of temperature (low) and concentration (high).

$$\begin{array}{l} H \quad O \\ | \quad \| \\ H-C-O-C-(CH_2)_{16}CH_3 \\ | \\ H-C-OH \\ | \\ H-C-O(CH_2CH_2O)_nH \\ | \\ H \end{array}$$

Ethoxylated monoglyceride (EMG)

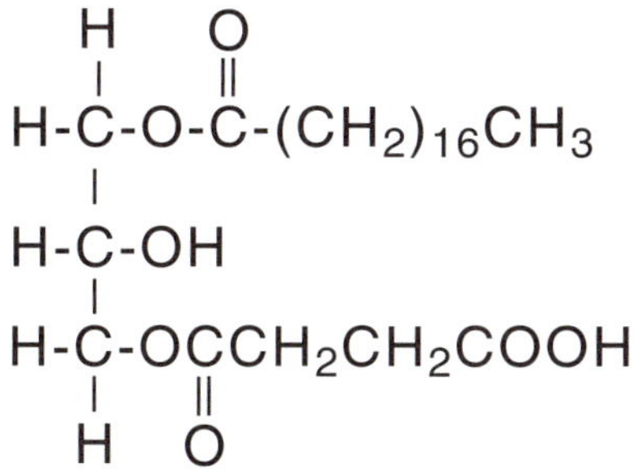

$$\begin{array}{l} H \quad O \\ | \quad \| \\ H-C-O-C-(CH_2)_{16}CH_3 \\ | \\ H-C-OH \\ | \\ H-C-OCCH_2CH_2COOH \\ | \quad \| \\ H \quad O \end{array}$$

Succinyl monoglyceride (SMG)

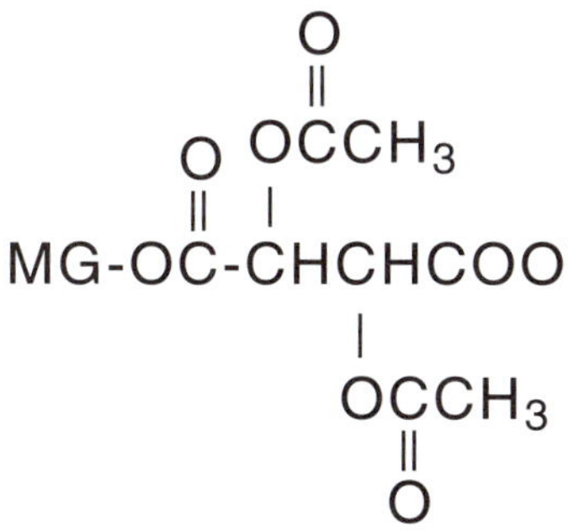

$$\begin{array}{l} \quad\quad\quad O \\ \quad\quad\quad \| \\ \quad O \quad OCCH_3 \\ \quad \| \quad | \\ MG-OC-CHCHCOO^- \\ \quad\quad\quad | \\ \quad\quad\quad OCCH_3 \\ \quad\quad\quad \| \\ \quad\quad\quad O \end{array}$$

Diacetyl tartrate ester of monoglyceride (DATEM)

Fig. 3-2. Emulsifiers used as dough strengtheners. In the DATEM structure, MG = monoglyceride.

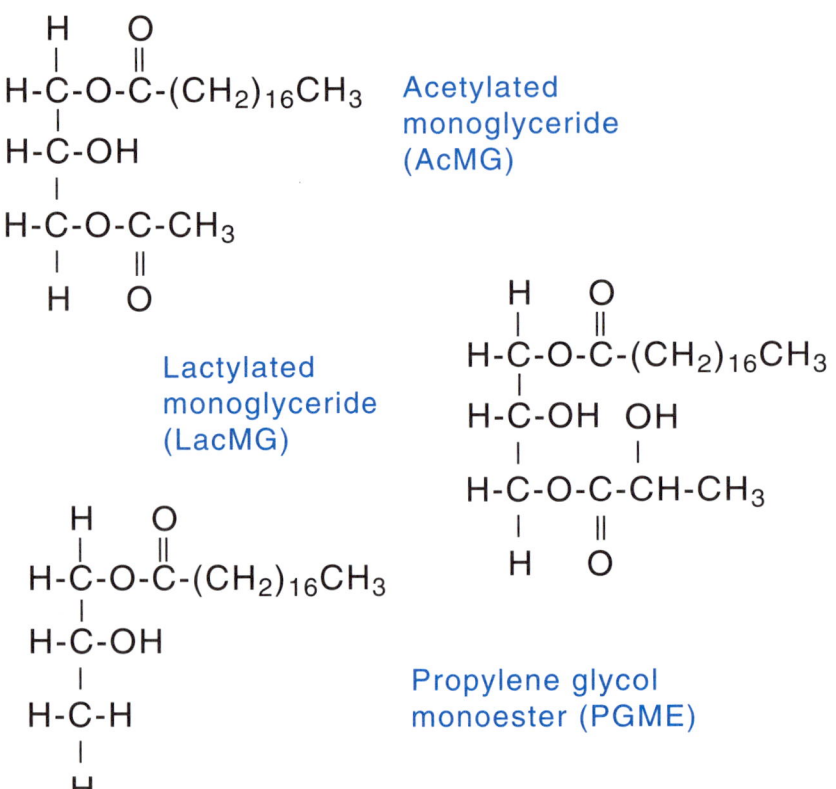

Fig. 3-3. α-Tending emulsifiers. These emulsifiers form solid films at the oil-water interface under conditions of low temperature and high concentration.

Primary hydroxyl—Hydroxyl group on the terminal carbon atom of a compound. That carbon atom is connected to only one other carbon.

Secondary hydroxyl—Hydroxyl group on a carbon atom that is attached to two other carbons.

Saponification value—Weight in milligrams of the potassium hydroxide required to saponify 1 g of a lipid. It characterizes a lipid by quantifying the proportion of ester groups relative to the total molecular weight.

propylene glycol monoester [PGME]). In each case, the structure shown is the main, but not the only, component of the mixture. The *primary hydroxyl* (3-hydroxyl) of monoglyceride is chemically more reactive than the *secondary* (2 position) *hydroxyl*. Thus, acetylation gives mainly 3-acetyl 1-monoglyceride, but there is also a certain amount of 2-acetyl 1-monoglyceride plus 1-acetyl 2-monoglyceride. In addition, any diglyceride present can also be acetylated.

SMG is made by reacting succinic anhydride with monoglyceride. The reaction is fairly straightforward. The various positional isomers mentioned for acetylated monoglyceride can also be found in succinylated monoglyceride.

DATEM is produced in two steps. First, tartaric acid (a byproduct of wine making) is reacted with acetic anhydride, acetylating the two hydroxyl groups and converting the tartaric acid to an anhydride. This is then reacted with monoglyceride. The major product is the structure shown in Figure 3-2, although numerous other closely related compounds are present. The four main reaction products are mono-(diacetyl tartaric acid) ester of monoglyceride, di-(diacetyl tartaric acid) ester of monoglyceride, mono-(diacetyl tartaric acid) ester of diglyceride, and mono-(diacetyl tartaric acid) monoacetyl ester of monoglyceride. The ratio of these four components in a given DATEM preparation depends mainly on the content of diacetyl tartaric anhydride and the type of monoglyceride used in the reaction. The functionality of the emulsifier depends to a large extent on its composition, so it is necessary to use a rather stringent raw material specification in order to ensure that performance remains consistent from lot to lot. Important indicator properties are *saponification value* and *acid value*.

The structure of EMG is even more random. Monoglyceride is treated with ethylene oxide gas under pressure in the presence of alkaline catalyst and at elevated temperatures. Ethylene oxide is polymerized via a series of ether linkages and also forms ether bonds with the free hydroxyl groups on monoglyceride. The average chain length is about 20 units ($n = 20$ in Fig. 3-2). Both the number 2 (β) and number 3 (α) positions of the monoglyceride may be derivatized, although because of its greater chemical reactivity, the primary hydroxyl (the α position) is more likely to be derivatized than the secondary hydroxyl (the β position). The exact distribution of polymer chain lengths and distribution between α and β positions are functions of reaction conditions, e.g., catalyst type and concentration, gas pressure, temperature, agitation, and length of reaction time. Of course, any diglyceride present may also be ethoxylated.

The second group of monoglyceride derivatives, the α-tending emulsifiers, are used mainly in cake production. These emulsifiers are dissolved in the shortening phase of the cake formulation, and they contribute to the emulsification of the shortening in the water phase and promote incorporation of air into the fat phase. The particular property of these emulsifiers that makes them valuable in liquid shortening cakes is that they form a solid film at the oil-water interface. This stabilizes the emulsion; but more importantly, it prevents the lipid phase from destabilizing the protein-stabilized foam during cake batter mixing (air incorporation).

The production of AcMG and PGME is straightforward organic chemistry. Treatment of monoglyceride with acetic anhydride results in the acetylated product, with the various kinds of isomers listed above. PGME can be made either by direct esterification of propylene glycol with fatty acids or by interesterification of fat (triglycerides) with propylene glycol. The direct esterification product typically contains about 55–60% monoester, and the remainder is diester. While a product containing >90% monoester is made by molecular distillation, the extra cost of this process is not warranted for most commercial (cake mix production) uses. The interesterified product is more complex, containing not only mono- and diesters of propylene glycol, but also about 10–15% monoglyceride and a small amount of diglyceride. As with DATEM, because such a wide range of product compositions is possible from different manufacturing processes, it is advisable to have stringent raw material specifications for this ingredient.

Lactic acid esters of monoglyceride are usually made by reacting lactic acid with a distilled monoglyceride. The complication here is that lactic acid contains a hydroxyl group, and the fatty acid moiety may migrate. For example, if lactic acid is heated with 1-monostearin, the main initial product is 3-lactoyl-1-stearoyl glycerol. However, during the reaction, some portion of the stearic acid may migrate to the lactyl hydroxyl, resulting in glyceryl 3-(stearoyl)-lactylate. In addition, lactic acid can polymerize (form lactoyl lactic esters), and lactoyl dimers and trimers may also be present. Thus, the reaction product mixture from heating lactic acid with a monoglyceride is a complex mixture containing as many as 10 identifiable molecular species. Production parameters must be tightly controlled to obtain a product with consistent functionalities.

Acid value—Weight in milligrams of potassium hydroxide required to neutralize the titratable groups in 1 g of lipid. It characterizes a lipid by quantifying the proportion of titratable acidic groups.

Hydrophilic/lipophilic balance (HLB)—Ratio of an emulsifier's hydrophilic and lipophilic tendencies.

SORBITAN DERIVATIVES

When the sugar alcohol sorbitol is heated with stearic acid in the presence of a catalyst, two reactions occur: sorbitol cyclizes to form the five-membered sorbitan ring and the remaining primary hydroxyl group is esterified by the acid. The resulting sorbitan monostearate (Fig. 3-4) is oil soluble and has a rather low *hydrophilic/lipophilic balance (HLB)* value (HLB is discussed later in this chapter). It is an approved additive for food use in many countries. Other important esters are sorbitan monooleate and sorbitan tristearate. A petition has been filed for affirmation as generally recognized as safe (GRAS) status for sorbitan tristearate, but until it is acted upon by the U.S. Food and Drug Administration (FDA), sorbitan tristearate is being sold as a self-affirmed GRAS crystal modifier.

Any of the three esters may be reacted with ethylene oxide to obtain polyoxyethylene derivatives (Fig. 3-5). The monostearate derivative is known as Polysorbate 60, the tristearate is Polysorbate 65, and the monooleate is Polysorbate 80. The discussion above of EMG regarding the length and location of the polyoxyethylene chains applies to these compounds. The average number of oxyethylene monomers is 20 ($n = 20$), and in the case of the monoesters, chains may be located on more than one hydroxyl group of the sorbitan ring (with the triester, of course, only one hydroxyl group is available for derivatization).

Sorbitan monostearate is a good emulsifier for use in making icings with superior aeration, gloss, and stability characteristics. It is also used as part of the emulsifier system in whipped toppings and coffee whiteners. The polyoxyethylene derivatives have found more acceptance; the monostearate Polysorbate 60 is the most widely used of the group. At a level of 0.25% (flour basis), the ability of Polysorbate 60 to strengthen dough against mechanical shock is greater than that of

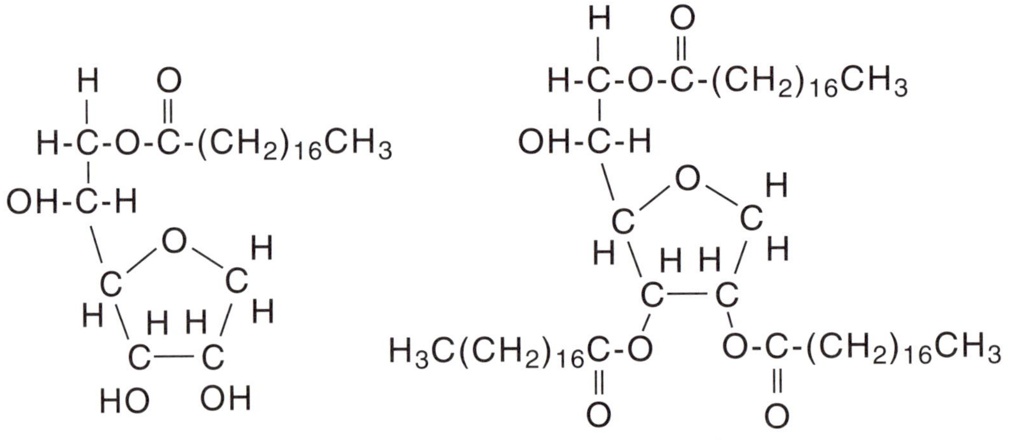

Sorbitan monostearate

Sorbitan tristearate

Fig. 3-4. Sorbitan esters. Note the mono- and tristearate esters of the sorbitan ring.

SMG and about equal to that of EMG and SSL. Polysorbate 60 has also been used in fluid oil cake shortening systems, generally in combination with GMS and propylene glycol monostearate.

POLYHYDRIC EMULSIFIERS

Polyglycerol esters. Polyglycerol esters (Fig. 3-6) have a variety of applications as emulsifiers in the food industry. The polyglycerol portion is synthesized by heating glycerol in the presence of an alkaline catalyst, and ether linkages are formed between the primary hydroxyls of glycerol. In the structure depicted in Figure 3-6, n may take any value; but for food emulsifiers, the most common are $n = 3$ (triglycerol), $n = 6$ (hexaglycerol), $n = 8$ (octaglycerol), and $n = 10$ (decaglycerol). (In all cases, n is an average value for the molecules present in the commercial preparation.) The polyglycerol backbone is then esterified to varying extents, either by direct reaction with a fatty acid or by interesterification with a triglyceride fat. Again, the number of

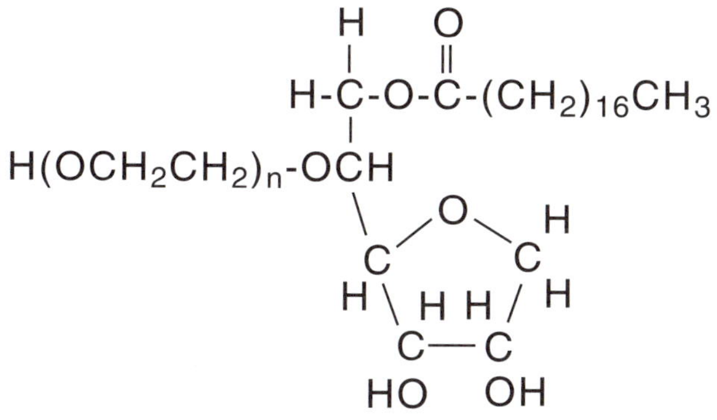

Polyoxyethylene (20) sorbitan monostearate (Polysorbate 60)

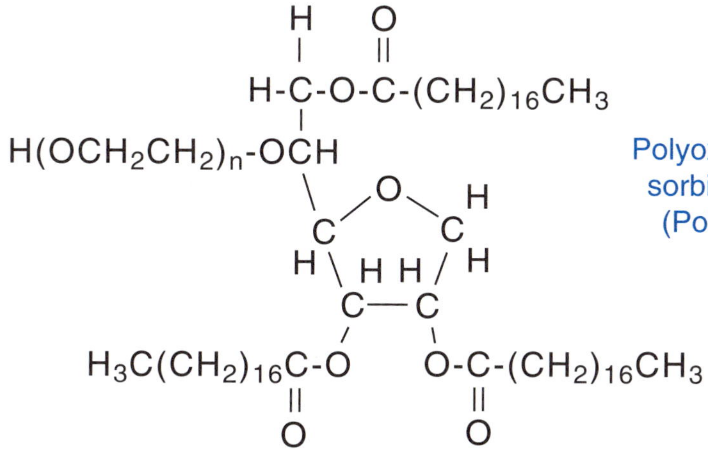

Polyoxyethylene (20) sorbitan tristearate (Polysorbate 65)

Fig. 3-5. Polysorbate emulsifiers. Reacting the sorbitan ester with ethylene oxide forms the hydrophilic polyoxyethylene side chain.

acid groups esterified to a polyglycerol molecule varies around some central value, so an octaglycerol octaoleate really should be understood as an (approximately octa)-glycerol (approximately octa)-oleate ester. By good control of feedstocks and reaction conditions, manufacturers manage to keep the properties of their various products relatively constant from batch to batch.

The HLB of these esters depends on the length of the polyglycerol chain (i.e., the number of hydrophilic hydroxyl groups present) and the degree of esterification. For example, decaglycerol monostearate has an HLB of 14.5, while triglycerol tristearate has an HLB of 3.6. Intermediate species have intermediate HLB values, and any desired value may be obtained by appropriate blending, as described below. The wide range of possible compositions and HLB values makes these materials versatile emulsifiers for food applications.

Sucrose esters. Sucrose has eight free hydroxyl groups, which are potential sites for esterification to fatty acids. Compounds containing six or more fatty acids per sucrose molecule are marketed as noncaloric fat substitutes under the name Olestra; this material acts like a triglyceride fat and has no surfactant properties. Compounds containing one to three fatty acid esters (Fig. 3-7), on the other hand, do act as emulsifiers and are approved for food use in that capacity. They are manufactured by the following steps:

1. An emulsion is made of fatty acid methyl ester in a concentrated aqueous sucrose solution.
2. The water is removed under vacuum at elevated temperature.
3. Alkaline catalyst is added, and the temperature of the dispersion is raised slowly to 150°C under vacuum, distilling off methanol formed during transesterification.
4. The reaction mixture is cooled and purified.

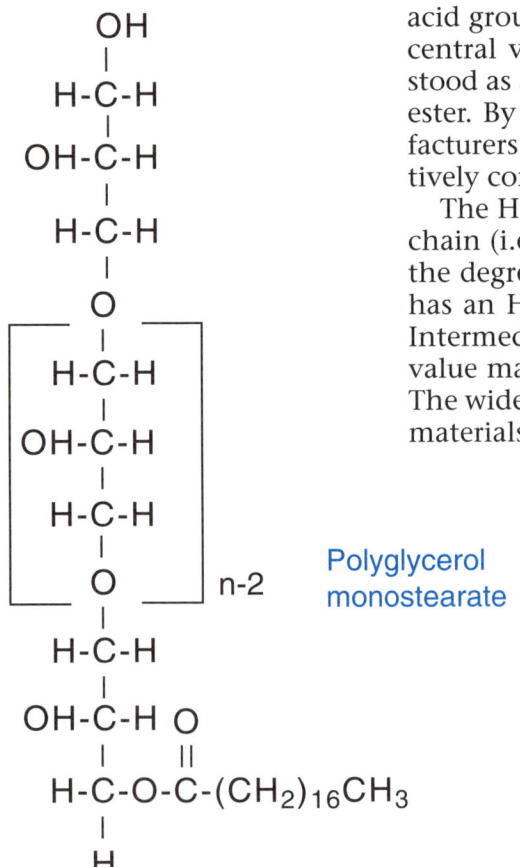

Fig. 3-6. Basic structure of the polyglycerol esters. The free hydroxyls on the polyglycerol can be esterified to fatty acids to varying extents, resulting in emulsifiers with a wide range of hydrophilic/lipophilic balance values.

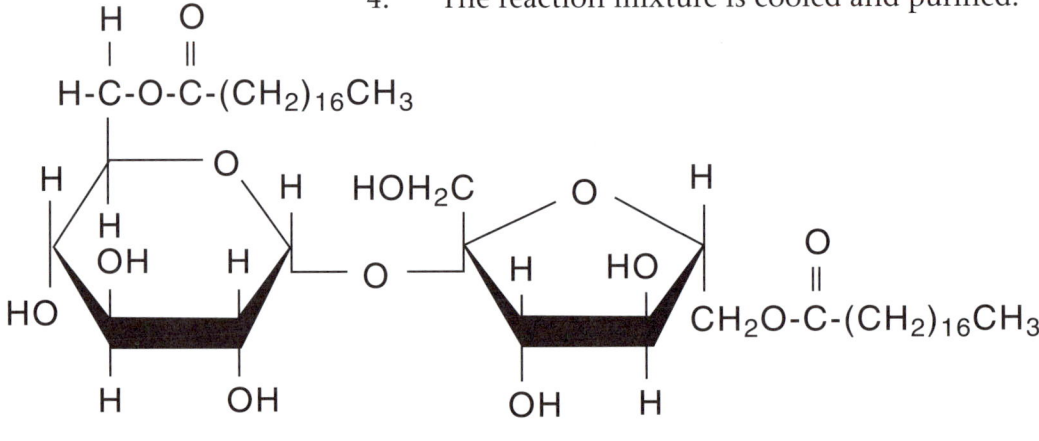

Fig. 3-7. One possible positional isomer of a sucrose diester.

TABLE 3-1. Ester Distribution in Sucrose Ester Emulsifiers

Designation	Percentage of Ester Type				HLB[a]
	Monoester	Diester	Triester	Tetraester	
F-160	71	24	5	0	15
F-140	61	30	8	1	13
F-110	50	36	12	2	11
F-90	46	39	13	2	9.5
F-70	42	42	14	2	8
F-50	33	49	16	2	6

[a] Hydrophilic/lipophilic balance.

The degree of esterification is controlled by the reaction conditions, especially the ratio of sucrose to methyl ester, and the final product is a mixture of esters (Table 3-1). The HLB value of a particular product is lower (more lipophilic) as the degree of esterification increases, as would be expected.

ANIONIC EMULSIFIERS

In addition to SMG and DATEM, some other anionic surfactants (Fig. 3-8) have been tried as dough strengtheners. SSL is currently the one most widely used in the United States. Sodium stearyl fumarate did not find acceptance, and sodium lauryl sulfate (SLS) is used mainly as a whipping agent with egg whites.

Lactic acid, with both a carboxylic acid and a hydroxyl function on the same molecule, readily forms an ester with itself. In commercial concentrated solutions, almost all the acid is present in this polylactylic form, and to get free lactic acid, it must be diluted with water and refluxed for a period of time. When stearic acid is heated with polylactic acid under the proper reaction conditions and then neutralized with sodium hydroxide, a product with the structure

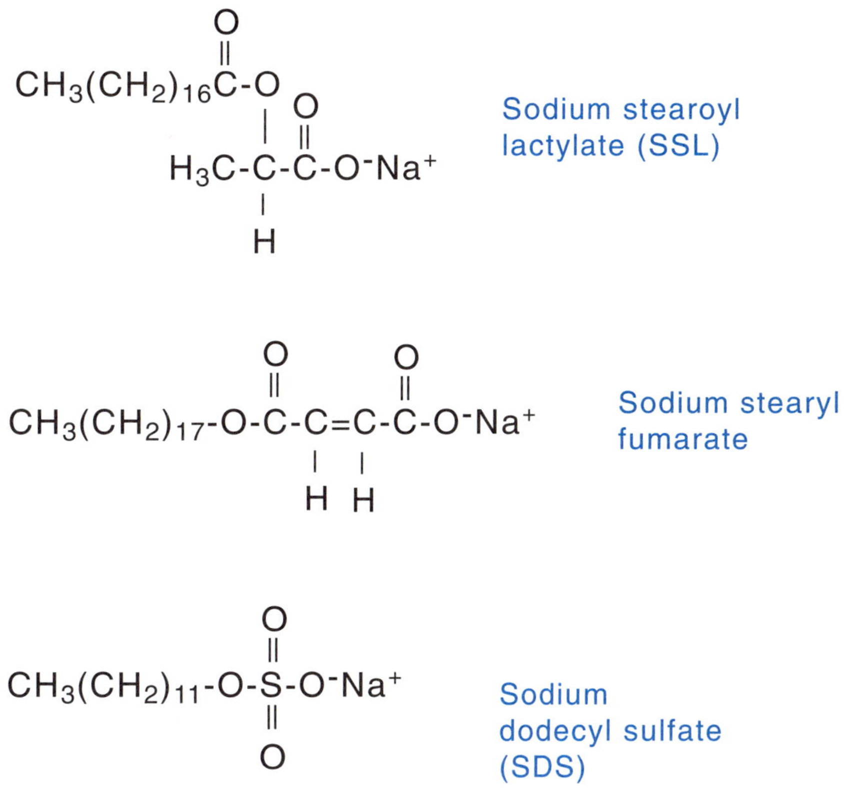

Fig. 3-8. Three anionic emulsifiers. SSL and the fumarate are dough conditioners, and SDS is used to solubilize proteins.

shown in Figure 3-8, is obtained. The monomer lactylic acid shown represents the predominant product; the dilactylic dimer is also present as well as lactylic trimers and tetramers. As with all compounds based on commercial stearic acid derived from hydrogenated fats, some percentage of the fatty acid is palmitic and small amounts of myristic and arachidic acids are also present. SSL is readily water soluble, but the calcium salt is practically insoluble. In this respect, it mimics a soap; e.g., sodium stearate is water soluble, but calcium stearate is oil soluble. Either form may be used, depending on the details of the intended application, but as a dough strengthener, the sodium salt is more common. As a stabilizer for hydrated monoglyceride, the sodium form is used because ionization in the water layer is necessary.

Stearyl fumarate is a half ester of fumaric acid with stearyl alcohol (octadecanol). Although stearyl fumarate might be expected to have dough-strengthening properties similar to those of SSL, this was not found to be so in practice, and the product was not a commercial success. Stearyl fumarate is still approved by the FDA for use in bread.

The third structure shown in Figure 3-8 is sodium dodecyl sulfate (SDS), often used by research workers for solubilizing proteins. It is a sulfate ester of the C12 alcohol dodecanol. The relationship between the pure compound (SDS) and the commercial surfactant (SLS) is as follows. Reduction of coconut oil yields a mixture called lauryl alcohol (from lauric acid, the predominant fatty acid in coconut oil). Sulfation and neutralization result in SLS. The alcohol portion of SLS is a mixture of chain lengths; the approximate composition of the common commercial form is 8% C8, 7% C10, 48% C12, 20% C14, 10% C16, and small amounts of longer chains. The most common food use of SLS is as a whipping aid. The compound is added to liquid egg whites at a maximum concentration of 0.0125% or to egg white solids at a level of 0.1%. It promotes the unfolding of *egg albumin* at the air-water interface and stabilization of the foam.

LECITHIN

The lecithin generally used by food processors is a by-product of the processing of crude soybean oil; it is the "gum" that is removed during the degumming step of oil refining. The *crude gum* is treated and purified to give the various commercial lecithin products that are available today. Crude soybean oil contains about 2% lecithin. Crude corn and cottonseed oils contain about 1% lecithin, but because smaller amounts of these oils are processed in the United States (compared with soy oil), the amount of gum obtained is usually too small for economical processing for human food uses. Instead, it is added back to animal feed formulations as a valuable source of energy. Egg yolk contains about 20% *phospholipid*, which accounts for its excellent emulsifying functionality, for example, in mayonnaise. However, isolated lecithin from egg yolk is too expensive to be used for food manufacture.

The crude gum is dehydrated (to remove water used during degumming), and then insoluble fines are removed by filtration. The crude

Egg albumin—Soluble protein found in egg white.

Crude gum—Material removed during the degumming phase of vegetable oil refining. Water is added to the crude oil, and the polar components (such as phospholipids) become hydrated and associated with the aqueous phase, which is then separated by centrifugation.

Phospholipid—Diglyceride esterified at the 3 position to phosphoric acid, which in turn is often esterified to another group.

material is brown to dark brown (depending on the amount of heat applied during processing) and contains some pigments extracted from the original soybean. It is bleached to attain a more acceptable light brown color. Treatment with up to 1.5% hydrogen peroxide results in a product known as single-bleached lecithin, and addition of benzoyl peroxide up to 0.5% yields double-bleached lecithin. Reaction with hydrogen peroxide at even higher levels plus lactic acid hydroxylates unsaturated fatty acid side chains at the double bond (e.g., yielding dihydroxystearic acid from oleic acid). The hydroxylated lecithin that is formed is more dispersible in cold water than the other types and is more effective as an emulsifier for oil/water (O/W) emulsions.

Phospholipids are insoluble in acetone, and the phospholipid content of lecithin is specified as *acetone insolubles* (AI). The standard commercial lecithin has a minimum AI content of 65%. Crude bleached lecithin is quite viscous. The addition of vegetable oil fluidizes lecithin, and commercial fluid lecithin products are standardized to have a viscosity of 7,500–10,000 cP at 25°C. A fully deoiled lecithin, a granular, free-flowing product with a typical AI content of 95–98%, is also produced.

The structures of the main surface-active components of lecithin are shown in Figure 3-9. The phosphatidyl group is a phosphate ester

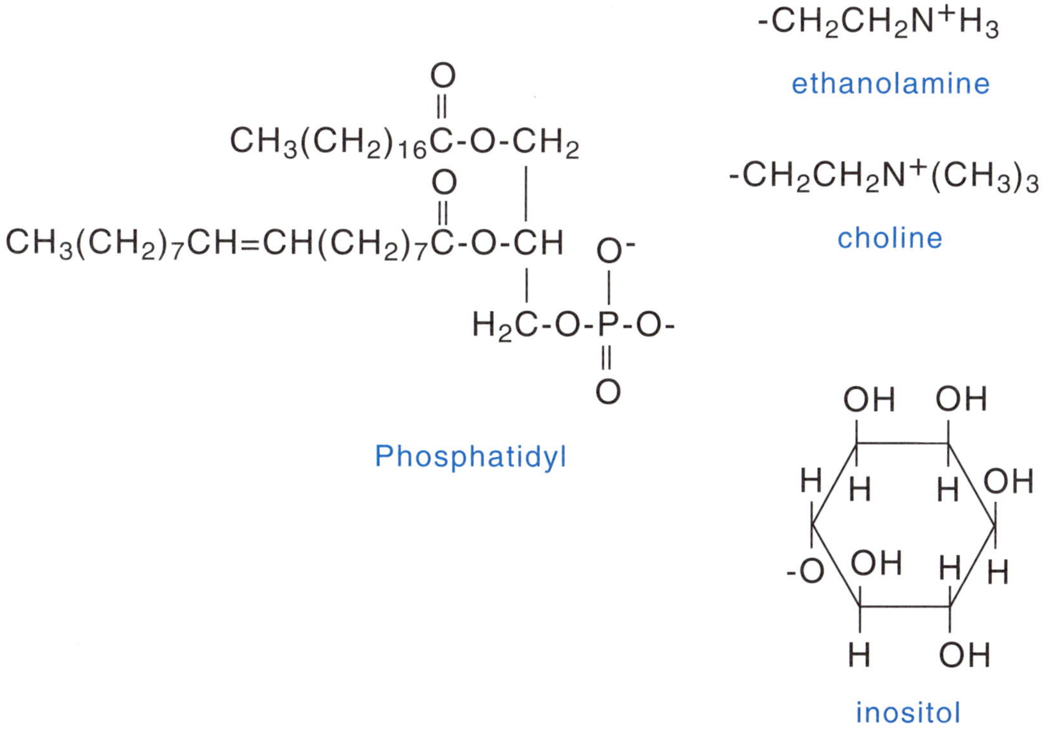

> **Acetone insolubles**—Specification of the amount of phospholipids in "gums," based on the fact that the other constituents normally present are soluble in acetone.

Fig. 3-9. Structure of the major components of soy lecithin. Serine is another possible substituent, but phosphatidylserine is found mainly in animal phospholipids, such as that from egg yolk. Removing one fatty acid from the phosphatidyl moiety results in a lysophospholipid.

of diglyceride. The fatty acid composition of the diglyceride is similar to that of the basic oil, so a number of different fatty acids are found, not just the stearic and oleic acids depicted. The three species are found in approximately equal amounts. Phosphatidylethanolamine (PE) and phosphatidylcholine (PC) are amphoteric surfactants, while phosphatidylinositol (PI) is anionic. Other surface-active materials are also found at somewhat lower concentrations. These include phosphatidic acid (the phosphatidyl moiety plus a hydrogen atom), lysophosphatides (the above species but with one fatty acid removed), and *glycolipids* (a sugar residue, either galactose or digalactose, attached to the free hydroxyl of a diglyceride).

The HLB values of the three types vary: PC has a high HLB, PE an intermediate value, and PI a low value. The HLB of the natural blend is 9–10, and emulsifier mixtures with values in this range tend to form either O/W or water/oil (W/O) emulsions, although neither type is highly stable. However, emulsifiers with intermediate HLB values are excellent wetting agents, and this is a major application for lecithin.

The emulsifying properties of lecithin can be improved by ethanol fractionation. PC is soluble in ethanol, PI is rather insoluble, and PE is partially soluble. Adding deoiled lecithin to ethanol results in a soluble and an insoluble fraction. The phosphatide composition of the ethanol-soluble fraction is 60% PC, 30% PE, and 2% PI plus glycolipids, and that of the ethanol-insoluble fraction is 4% PC, 29% PE, and 55% PI plus glycolipids. The remaining small percentage of both fractions includes oil, free fatty acids, and lysophosphatides. The soluble fraction is effective in promoting and stabilizing O/W emulsions, while the insoluble portion promotes and stabilizes W/O emulsions. Chromatography of the ethanol-soluble fraction can be performed to obtain a material containing 90% PC that is used to make an egg yolk replacer. The emulsifying ability of egg yolk results in part from its high HLB (70% PC and 15% PE in the total yolk phospholipid) plus the protein. The fractionation process is currently used by several European companies to produce industrial food-grade emulsifiers with special functionalities.

Hydrophilic/Lipophilic Balance

BASIC PRINCIPLE OF THE CONCEPT

As discussed in Chapter 1, emulsifiers consist of a hydrophilic portion (consisting of a wide variety of structures) and a lipophilic portion (usually a fatty acid or occasionally a fatty alcohol). The balance between these two governs the functionality of the emulsifier at interfaces and hence its utility in foods. This balance, called the hydrophilic/lipophilic balance (HLB), attaches a number to emulsifiers that guides the food technologist in choosing one for a particular application.

The initial proposal was to calculate HLB as

$$HLB = L/T \times 20$$

Glycolipid—Diglyceride connected to a sugar moiety (usually galactose or galactosylgalactose) at the 3 position and common in cereal and legume seeds.

in which L is the molecular weight of the hydrophilic part of the molecule and T is the total molecular weight. Thus, a pure hydrocarbon would have an HLB of 0, while a pure hydrophile (e.g., sugar) would have an HLB of 20.

Another method of calculating HLB is to add up contributions by various functional groups. The functional groups and their associated group numbers are listed in Table 3-2. The group values for the hydrophilic and lipophilic functions in the emulsifier are summed and inserted into the equation:

HLB = Σ (hydrophilic values) − Σ (lipophilic values) + 7

The agreement between calculated and experimentally determined HLB values is generally within a few tenths. Considering that neither method is highly precise, one cannot expect much better, and the calculated value gives a good indication of HLB for a new emulsifier under investigation.

HLB values provide a guide to the functionality of the emulsifier system. The following guidelines are based on experience:

- HLB of 3–6: a good water-in-oil emulsifier;
- HLB of 7–9: a good wetting agent; and
- HLB of 10–18: a good oil-in-water emulsifier.

TABLE 3-2. Hydrophilic/Lipophilic Balance Functional Group Numbers

Group	Group Number
Hydrophilic	
$-SO_4Na$	38.7
-COOK	21.1
-COONa	19.1
Sulfonate	~11
$-N(CH_3)_3$	9.4
Ester (sorbitan ring)	6.8
Ester (others)	2.4
-COOH	1.9
-OH (sorbitan ring)	0.5
-OH (other)	1.9
$-(CH_2-CH_2-O)-$	0.33
Lipophilic	
-CH-	...
$-CH_2-$	0.475
$-CH_3$	...
=CH-	...

EXPERIMENTAL DETERMINATION OF HLB

While theoretical calculations are fine, emulsifiers are used in food systems that, for the most part, are anything but theoretically simple. For practical purposes, determination of optimum HLB for a particular system is best done experimentally. This approach starts with the concept that the HLB of a blend of emulsifiers is the algebraic sum of their contributions. For example, a blend of sorbitan monostearate (HLB = 4.7) and Polysorbate 60 (HLB = 14.9) is used to determine the optimum HLB for a salad dressing formulation. A 50/50 mixture of the two has an HLB of 9.8 (0.5 × 4.7 + 0.5 × 14.9), a 25/75 blend has an HLB of 12.35 (0.25 × 4.7 + 0.75 × 14.9), etc. While this is a straight-line relationship, it should be noted that in practice the prediction does not hold precisely true.

The first step is to prepare a series of emulsifier blends at intervals of 0.5 HLB units. The range should be selected on the basis of the desired product characteristics (W/O or O/W emulsion) and according to the guidelines above. An excess of emulsifier (approximately 10% of the weight of the oil phase) is used to ensure emulsion formation. The emulsifier is dissolved in the oil, the aqueous components are added, and emulsions are made for each trial by using a standardized agitation technique. The emulsions are allowed to stand and are then

Response surface methodology—Method of experimental design to determine the optimum values for two or more variables in a product by using a limited number of experiments and making interpolations based on the experimental data.

assessed for stability, e.g., by measuring the thickness of the oil (or water) layer formed at various times. (If all the emulsions are too stable to break down in a reasonable length of time, the experiment should be repeated with less emulsifier.) The result is an approximate optimum HLB range for the system.

The next step is to choose the best chemical type of emulsifiers to use. Experience shows that a blend of two emulsifiers (one lipophilic and one hydrophilic) generally produces the most stable emulsions. Numerous such pairs are available, e.g., sorbitan monostearate plus a polysorbate, monoglyceride plus a fatty acid salt (sodium or potassium), triglycerol tristearate plus decaglycerol monostearate, and sucrose esters (e.g., F-160 plus F-50).

In addition, the nature of the fatty acid chain (e.g., chain length and degree of unsaturation) can sometimes make a difference that is readily seen by comparing some of the entries in Table 3-3. The melting point can be a factor in some processing systems; solid monostearin may be difficult to add, whereas liquid monoolein (with the same HLB) may be more convenient to use. If the system is acidic (e.g., a salad dressing), acid-stable emulsifiers such as the sorbitans or polyglycerols may be the best choices. If the pH is in the neutral range, the monoglyceride plus fatty acid salt pair may be the most effective.

The various combinations are then subjected to emulsifying trials as described above, but blends that are in the middle of the apparent optimum HLB range are used. These trials should include relatively small HLB variations (i.e., optimum and ±0.5 HLB units) and limited variation in concentrations. The goal is to choose the emulsifier system that provides the best stability at the lowest usage level.

After the type of emulsifier is selected, the system can be fine-tuned by running a series of trials at intervals of 0.1 HLB units and at various concentrations to find the minimum amount that yields the desired emulsion stability. At this stage of testing, using *response surface methodology* maximizes information gained while minimizing the amount of work needed.

Two final points must be emphasized. First, HLB is an empirical system for characterizing emulsifiers, and the values assigned to emulsifiers (as in Table 3-3) are necessarily somewhat imprecise. Thus, if a particular emulsifier combination provides the best results at an HLB of 11.4, a different combination might function best at an HLB of 11.7. The effects of any substitutions must be evaluated, although the use of the HLB system greatly decreases the amount of work needed. Second, emulsifiers interact with other food ingredients, and if formula changes are made (additions or subtractions),

TABLE 3-3. HLB[a] Values of Food Emulsifiers

Emulsifier	Experimental HLB
Sodium lauryl sulfate	40
Sodium stearoyl lactylate	22
Potassium oleate	20
Sucrose monoester	20
Sodium oleate	18
Polysorbate 60	15
Polysorbate 80	15
Decaglycerol monooleate	14
Decaglycerol monostearate	13
Ethoxylated monoglyceride	13
Decaglycerol dioleate	12
Polysorbate 65	11
Hexaglycerol dioleate	9
Decaglycerol hexaoleate	7
Triglycerol monostearate	7
Glycerol monolaurate	7
Sorbitan monostearate	5.9
Sucrose triester	5
Propylene glycol monolaurate	4.5
Propylene glycol monostearate	3.4
Glycerol monostearate	3.8
Sorbitan tristearate	2.1

[a] Hydrophilic/lipophilic balance.

the assessment of optimum HLB must be repeated. This is especially true if the changes involve proteins or gums.

Proteins

Proteins are surface-active molecules. They unfold (i.e., denature, or lose their native three-dimensional structure) at water-air or water-oil interfaces, adsorbing at the interface and stabilizing the foam or emulsion. Their contribution to foam stability is briefly discussed in Chapter 1, and the properties and forces contributing to unfolding and adsorption are discussed here.

FOAMING AGENTS

The best-known example of proteins as foam stabilizers is the whipping of egg white to make meringue. Ovalbumin (the main protein in egg white) readily denatures and spreads at the air-water interface. When whipping is continued to the "dry-peak" stage, all the water is immobilized in the interstitial spaces and the meringue does not drain, even after an extended period of time. Air bubble size has been reduced (and numbers increased), generating enough interfacial area to accommodate all the protein in the unfolded state. Many other soluble proteins can also be whipped, but the resulting foam is less stable. Drainage occurs, and after a period of time, the foam begins to break down. The difference may be attributed to reversibility of the unfolding process. For ovalbumin, the process is essentially irreversible, and it stabilizes the interface, even though system energy is high. Soluble soy protein unfolds, but with time (and under the energetic influence to lower free energy by diminishing interfacial area), the adsorbed protein is slowly forced back into the aqueous phase, causing the foam to collapse. Acetylation of soy protein (converting basically charged lysine amino groups into uncharged acetamido groups) provides a soluble protein in which the resolubilizing tendency is greatly reduced, and stable, dry-peak foams can be made from such a protein.

In cake batters, the soluble flour proteins (and, of course, any added egg whites) act as foaming agents. In cakes made with oil, proteins are the main agents of air incorporation. The destabilization of foams in these cakes must be prevented by using an α-tending emulsifier that encapsulates the oil (see Chapter 4). In cakes made with fractionated and reconstituted flours, omission of the soluble protein fraction results in failures (1).

EMULSIFYING AGENTS

Proteins at interfaces. A protein is a *polypeptide* (i.e., a chain of amino acids joined by amide linkages) in which some of the amino acid side chains are hydrophilic and some are hydrophobic. The chain is coiled upon itself in such a way that most of the hydrophobic side chains are

Polypeptide—A polymer consisting of amino acids connected by an amide bond, involving the carboxylic acid and α amino groups.

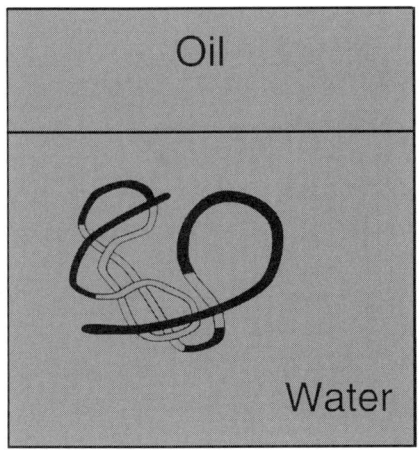 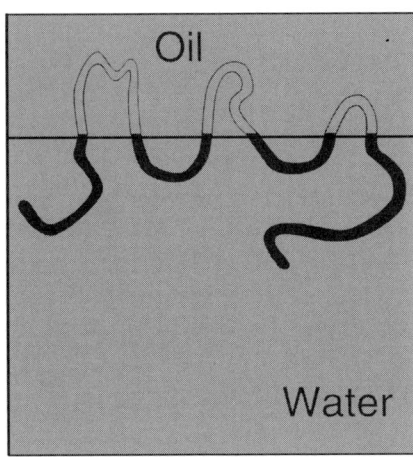

Fig. 3-10. Left, protein dissolved in water, folded into its native configuration. Right, protein unfolded at an interface. The hydrophobic parts of the chain (light gray sections) penetrate the oil side of the interface while the hydrophilic parts (black sections) remain in the water.

Complexation—Combination of two different molecular species.

in the interior of the protein (and not in contact with water) and the hydrophilic chains are on the exterior surface where they can interact with water (Fig. 3-10, left). A protein has been characterized as "an oil drop surrounded by a hydrated shell." This conformation is not rigid, and under the right conditions it can unfold. If an interface is present, the hydrophobic parts of the chain will penetrate the air (or oil) side of the interface while the hydrophilic parts remain in the water (Fig. 3-10, right). The protein acts as an amphiphile, and the surface (or interfacial) tension is lowered.

Both soluble and insoluble proteins can emulsify oil. Myosin, a low molecular weight protein related to hemoglobin, emulsifies fat during the grinding of meat to make sausage. Other proteins such as isolated soy protein are also often added to the mix, emulsifying fat (and increasing its contribution to sausage net weight) as well as holding water during cooking (thereby increasing yield).

The lipoproteins in egg yolk are the main emulsifying agents in mayonnaise (see Chapter 6). Lipoproteins from other sources (e.g., peanut flour) can also emulsify oil, and a type of mayonnaise has been made with such preparations. The emulsification capabilities of egg yolk lipoproteins are enhanced by the lower pH that results from the vinegar and/or lemon juice used in such dressings.

A major difference between proteins and ordinary emulsifiers (e.g., a sodium soap) relates to size. The molecular weight of milk proteins is 20,000–300,000, while that of sodium oleate is 307. Thus, soap diffuses to the interfacial region much more quickly than a protein. When the protein reaches the interface, the unfolding of the chains and partitioning of the segments between the two phases also take time. If oil is added to a dilute soap solution and to a dilute protein solution, the former system reaches its final interfacial tension value within a matter of seconds, while the latter interface may take up to an hour to reach final equilibrium. Temperature, pH, and different types and concentrations of salts also greatly affect the time required for the protein to complete its adsorption at the interface.

Protein-emulsifier interactions. Protein and emulsifier molecules can interact in two ways: *complexation*, which is usually observed when both types of molecules are in solution, and competition, which represents the interaction at interfaces.

Since a protein is partially hydrophobic in nature, it is not surprising that it can interact with the hydrophobic portion of an emulsifier. The binding of amphiphiles such as SDS or fatty acid to blood serum albumin has been studied in great detail, in part because the protein is readily available in a pure state, allowing easier interpretation of experimental data. A more practical example is the complexation of a dough strengthener such as SSL to gluten (see Chapter 4).

At low levels of interaction, up to about 10 moles of emulsifier per mole of protein, there is no alteration in protein configuration, but as the binding ratio increases, the protein begins to lose its native configuration. As the amount of complexed emulsifier increases, the amount of the hydrophobic component of the emulsifier becomes significant compared with the amount of hydrophobic region in the protein, and the influence of amino acid side chains in maintaining the native protein configuration (through hydrophobic bonding) is lessened. The influence of the ionic charge of the emulsifier hydrophilic group is much more disruptive. As the binding ratio approaches 100, the negative charges contributed by the emulsifier (assuming it is SDS or a soap) overcome the charged amino acid side chains, and the protein-emulsifier complex takes on a fairly high charge density. The ionic interactions (e.g., carboxylate-amino group pairs) that contribute to the stability of the native protein configuration are disrupted, and the protein unfolds. The unfolded complex has been called the "necklace on a string" model (Fig. 3-11). Micelles of the emulsifier incorporate hydrophobic regions of the protein. Since these micelles are all negatively charged, they are mutually repulsive, and the necklace is rod shaped.

Box 3-1. SDS-PAGE

The ability of 1% SDS in buffer to convert most proteins to rod-shaped molecules with lengths proportional to their molecular weights has been used to estimate protein size by polyacrylamide gel electrophoresis (PAGE). The gel is a spongy structure containing buffer. A sample of SDS-denatured protein is placed at one end of the gel slab, and an electrical potential is applied. The complex migrates toward the anode (positively charged pole). The rate at which the complex makes its way through the gel is proportional to its length—small rods travel faster than large ones. After an appropriate time, the gel slab is removed from the electrophoretic apparatus, and a protein stain is applied so that the extent of migration can be measured. This is compared with the migration rates of a standard mixture of proteins of known molecular weight (run in the gel at the same time), and the molecular weight of the unknown protein can be estimated.

The binding described occurs only above the critical micelle concentration (CMC) of the emulsifier (see Chapter 2). The binding mode is termed "cooperative"; i.e., the initial binding (and unfolding of a small region of the protein) enhances further binding and unfolding. Nonionic emulsifiers complex with proteins, although the effect is small, since in general the emulsifiers are relatively insoluble in water. The interaction of a moderately soluble emulsifier, octyl glucoside, with many proteins has been studied. Cooperative binding of more than 100 molecules of the emulsifier per molecule of protein was observed. However, there was no measurable disruption of protein configuration, which highlights the importance of the charge disruption to the unfolding described above.

As mentioned above, proteins can unfold and adsorb at interfaces (either air-water or oil-water). If another emulsifier is also present in the system, it too adsorbs at the interface. Several reactions are possible.

- Competitive adsorption: The two surface-active molecules compete for the available interfacial area.

- Displacement: The more surface-active compound displaces the less active material from the interface.

- Enhancement: A complex of the small molecule with the protein increases the interfacial action of the protein.

- Reinforcement: Interaction at the interface results in more efficient packing of the elements involved (the hydrophobic regions of the protein and the hydrophilic ends of the emulsifier), increasing the total interfacial concentration and decreasing interfacial tension even further.

Numerous permutations and combinations of these possibilities have been observed, mainly at air-water interfaces but also occasionally at oil-water interfaces. The proteins studied have ranged from purified (e.g., ovalbumin and lactoglobulin) to complex (e.g., gluten and meat extracts), and all types of emulsifiers have been examined, although most reports involve either nonionic or anionic food emulsifiers. A typical example of each case follows.

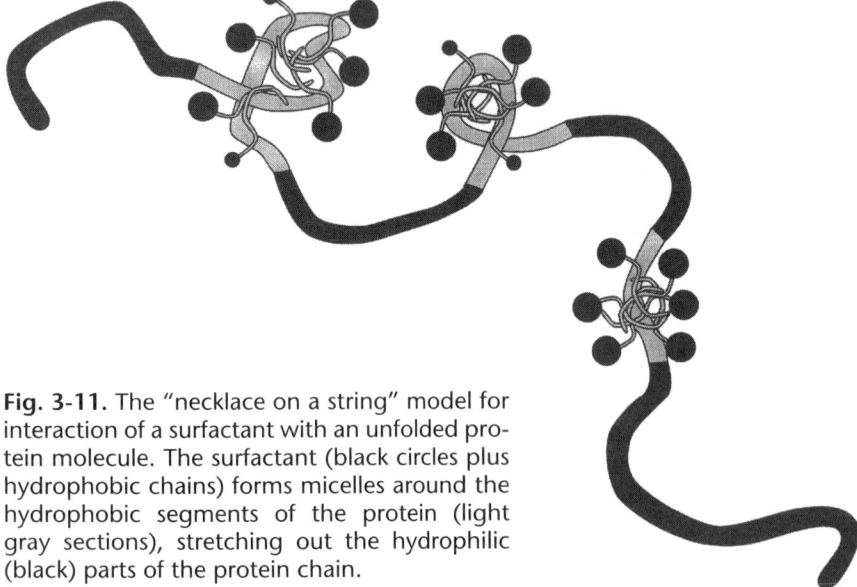

Fig. 3-11. The "necklace on a string" model for interaction of a surfactant with an unfolded protein molecule. The surfactant (black circles plus hydrophobic chains) forms micelles around the hydrophobic segments of the protein (light gray sections), stretching out the hydrophilic (black) parts of the protein chain.

At low concentration, proteins tend to be more surface active than emulsifiers. The addition of a mixture of protein (e.g., β-casein) plus a surfactant such as SDS (at a concentration below its CMC) results in a surface tension that is about the same as that of protein alone. Raising the surfactant concentration to near its CMC allows it to become the dominant surface-active molecule, and the surface tension then becomes the same as that of SDS alone.

In a drained protein foam, the unfolded protein stabilizes the film between air bubbles and prevents foam breakdown. If a small amount of a nonionic emulsifier (e.g., monocaprin) is present in the liquid, the foam is rather unstable and tends to collapse quickly after it drains. The small molecules lower surface tension (so the foam can be formed) but are not able to prevent the formation of holes in the film that lead to collapse.

If the water-soluble emulsifier binds to the protein, it may actually enhance the unfolding and surface-adsorption phenomenon. This is the case with SLS and egg white protein. The surfactant, present at low level, presumably destabilizes the protein conformation, leading to quicker adsorption at the air-water interface and better whippability of the egg white.

Surface tension decreases as SDS concentration increases up to the CMC of SDS and then remains constant (Fig. 1-3). In the presence of ovalbumin, for instance, the surface tension is less than that of SDS alone up to the CMC, but at concentrations above the CMC, surface tension continues to decrease. This must result from additional adsorption of the protein-SDS complex at the surface.

Most studies of surfactants include measurements of interfacial tension (see Chapter 1). However, adsorbed proteins usually show a large increase in *interfacial viscosity*, presumably because they are macromolecules. Often, this is the property that makes the protein functional in the food system of interest. A viscous film at the oil-water interfaces in an emulsion, for instance, greatly increases the resistance to coalescence of oil droplets. Certain surfactants may displace (or disrupt) the protein film, lowering interfacial viscosity while at the same time producing the same or lower interfacial tension. An emulsion made with protein alone is more stable (i.e., less coalescence is observed) than an emulsion made with the protein plus a small amount of the displacing surfactant.

These considerations play a role in many types of food systems: bakery batters, meat (sausage) emulsions, and egg yolk-stabilized dressings and sauces. The influence of protein at interfaces, however, is most pervasive in dairy products (see Chapter 5).

Regulations

Emulsifiers are additives to traditional foods, and as such their use is regulated by most governments. In the United States, they are approved by the FDA in one of three categories (approved additives,

Interfacial viscosity— Resistance to flow in the two dimensions of the interface.

GRAS materials, or affirmed as GRAS compounds) and listed in the *Code of Food Regulations*, Title 21. In Canada, emulsifiers that may be used are listed in Table IV, Division 16, of the *Canadian Food and Drug Regulations*. In Europe, the European Union issues European Parliament and Council Directives that list food additives authorized for use in human foodstuffs.

Most of the emulsifiers commonly used in foods are listed in Table 3-4. This table should be used only as a guide to the regulations. Many emulsifiers are approved only for certain kinds of food products, and often the allowable amount is limited. Before a large-scale product-development project is begun, the actual regulations concerning the emulsifiers to be used should be consulted to make sure that the final product meets all regulatory guidelines.

Reference

1. Howard, N. 1972. *Bakers Dig.* 46(5):28-30, 32, 34, 36-37, 64.

Supplemental Reading

Friberg, S. E., and Larsson, K., Eds. 1997. *Food Emulsions*, 3rd ed. Marcel Dekker, New York.

TABLE 3-4. Regulatory Status of Emulsifiers

Emulsifier	United States[a]	Canada[b]	European Union[c]
Mono- and diglycerides (GRAS)[d]	182.4505	M.4, M.5	E 471
Succinyl monoglyceride	172.830	...[e]	...
Acetylated monoglyceride	172.828	A.2	E 472a
Lactylated monoglyceride	172.852	L.1	E 472b
Monoglyceride citrate	172.832	...	E 472c
Monoglyceride phosphate (GRAS)	182.4521	A.94, C.7	...
Diacetyl tartrate ester of monoglyceride (GRAS)	182.4101	A.3	E 472e
Stearyl monoglyceride citrate	172.755	S.19	E 472f
Polyoxyethylene monoglyceride	172.834	...	...
Polyoxyethylene (8) stearate	...	P.5	...
Propylene glycol monoester	172.854	P.14	E 477
Lactylated propylene glycol monoester	172.850	...	...
Sodium, potassium salts of fatty acids	172.863	...	E 470
Sorbitan monostearate	172.842	S.18	E 491
Sorbitan tristearate	...	S.18B	E 492
Polysorbate 60	172.836	P.3	E 435
Polysorbate 65	172.838	P.4	E 436
Polysorbate 80	172.840	P.2	E 433
Calcium stearoyl lactylate	172.844	...	E 482
Sodium stearoyl lactylate	172.846	S.15A	E 481
Stearoyl lactylic acid	172.848	L.1A	...
Stearyl tartrate	...	...	E 483
Stearoyl propylene glycol hydrogen succinate (succistearin)	172.765	...	...
Sodium stearyl fumarate	172.826	...	...
Sodium lauryl sulfate	172.822	...	...
Dioctyl sodium sulfosuccinate	172.810	...	...
Polyglycerol fatty acid esters	172.854	P.1A	E 475
Sucrose fatty acid esters	172.859	S.20	E 473
Sucrose glycerides	...	...	E 474
Lecithin (GRAS)	184.1400	L.2	E 322
Hydroxylated lecithin	172.814	H.1	E 322
Oxystearin	172.818	...	...
Triethyl citrate (GRAS)	182.1911	...	...

[a] *Code of Food Regulations*, Title 21.
[b] *Canadian Food and Drug Regulations*, Table IV, Division 16.
[c] European Parliament and Council Directive 95/2/EC, 20 February 1995.
[d] Generally recognized as safe.
[e] Not listed by this jurisdiction.

CHAPTER 4

Bakery Products

Surfactants used in the production of bakery goods are usually referred to as either "emulsifiers" or "dough strengtheners." From a physical chemist's point of view, the way these terms are used is imprecise. The interfacial role of these materials is spelled out in this section, but in the remainder of this chapter, current bakery terminology is used for the sake of convenience.

Strictly speaking, an emulsifier is a surfactant that promotes the formation of an emulsion; that is, it aids in the subdivision of particles of the discontinuous phase. In bakery usage, this function is most important in the production of batters for cake, cake doughnuts, waffles, etc.

The term "emulsifier" is also applied to compounds (i.e., crumb softeners) that interact with molecules and granules of gelatinized starch and slow the rate at which they recrystallize, thereby contributing to the retention of crumb softness. Surfactants that perform this role react at the solid-liquid, not the liquid-liquid, interface.

Dough strengtheners are surfactants that presumably interact with gluten proteins and enhance the dough characteristics that bakers call "strength." Again, the functionality is at the solid-liquid interface.

All bakery surfactants aid the incorporation and subdivision of air into the liquid phase; that is, they promote foam formation. This is important in cake production and in the generation of fine-grained crumb in bread. Compounds used for this purpose are usually called emulsifiers, although they are actually foaming agents that function at the gas-liquid interface.

Finally, it should be noted that any given surfactant may function in all the ways listed, even though it is used primarily for one specific function. For example, sodium stearoyl lactylate (SSL) is used mainly for its dough-strengthening effect, but it also promotes emulsification, air incorporation and subdivision, and retention of crumb softness. Again, in the viewpoint of the physical chemist, a bakery dough or batter is a "messy" system with a multitude of interfacial interactions occurring simultaneously. It is helpful to isolate each type of interaction for discussion, realizing that ultimately all functions must be considered together to realistically assess the effects of surfactants on bakery foods.

In This Chapter:

Antistaling Agents
  Starch Gelatinization
  Starch Retrogradation
  Bread Staling
  Emulsifier-Starch
    Complexation

Dough Strengtheners

Aeration Agents

Troubleshooting

Antistaling Agents

Emulsifiers were first used in bread to extend shelf life, i.e., to retard staling. Monoglycerides are the primary antistaling additives used today and account for about one-third of the emulsifiers used in the baking industry. Staling is a complicated process that involves changes in all the components of bread and is actually a sensory response (older bread tastes "stale") to these changes. A decrease in flavor impact or a drier mouthfeel can be measured by taste panels, but the consumer commonly uses the "squeeze test." Staleness is equated with resistance to manual squeezing of a loaf of bread. Bakery quality-control laboratories and research groups usually assess staleness by measuring the resistance of the crumb to compression in an instrument such as the Baker Compressimeter, the Instron Universal Testing Machine, or the penetrometer. Crumb compression correlates with panel ratings; there is an excellent linear relationship between panel freshness score (which includes flavor and mouthfeel as well as texture) and the logarithm of the *elastic modulus*, or *crumb modulus* (resistance to compression).

From the extensive research done on staling, three conclusions are important:

1. Staling is not related to moisture loss from the bread. A five-day-old stale loaf (stored under the proper conditions) has the same moisture content as fresh bread, although it gives a drier mouthfeel impression.
2. Staling is related to the recrystallization (retrogradation) of the starch molecules gelatinized during the baking process.
3. Other bread components (e.g., gluten protein and pentosans) may play a role, but the extent and the nature of their contribution to staling is uncertain.

This section is concerned with the role emulsifiers (mainly monoglycerides) play in retarding bread staling. To begin the discussion, the current understanding of the involvement of starch in staling is first reviewed.

STARCH GELATINIZATION

Native *starch granules* are ordered structures possessing a high degree of crystallinity. They are made up of two types of polyglucoside chains: *amylose*, in which the linkages between monomers are of the α-1,4 type, and *amylopectin,* in which there are both α-1,4 and α-1,6 linkages. Amylose is a linear chain polymer with a molecular weight of 100,000–300,000. Amylopectin is a branched structure with a molecular weight of 10,000,000–40,000,000. Segments of several adjacent chains may order themselves in a crystalline structure (*crystallites*), while other parts of the molecules are more random. The internal structure of the starch granule is still under discussion, but it may be roughly characterized as containing both crystalline and

---

Elastic modulus—Relationship between stress (force) applied to a sample and the strain (deformation) in the sample; a more general rheological term than crumb modulus.

Crumb modulus—Synonym for elastic modulus.

Starch granules—Naturally occurring, partially crystalline, discrete aggregates of amylose and amylopectin.

Amylose—Linear polyglucose chains in starch.

Amylopectin—Branched polyglucose chains in starch.

Crystallites—Small regions of crystalline starch within a granule.

amorphous regions. The crystallinity is evidenced by X-ray diffraction patterns of native starch, by a sharp melting endotherm when starch granules are heated in water in a *differential scanning calorimeter (DSC)*, and by the *birefringence* (Maltese cross) seen when native starch is viewed microscopically in polarized light. When starch is heated in water, crystallinity begins to disappear at a temperature that varies depending on the source of the starch, and it is almost completely lost at a point about 6–8 degrees C higher. This loss of structure is called *gelatinization* (1):

> Starch gelatinization is the collapse (disruption) of molecular orders within the starch granule manifested in irreversible changes in properties such as granular swelling, native crystallite melting, loss of birefringence, and starch solubilization. The point [temperature] of initial gelatinization and the range over which it occurs is governed by starch concentration, method of observation, granule type, and heterogeneities within the granule population under observation.

When starch granules are heated in water, the first change noted microscopically is a gradual increase in granule volume, or swelling. Generally, the granules take up available water, and the loss of crystallinity is progressive. At intermediate stages of heating or in situations where water is limited (a starch-water ratio of two or more), some birefringence persists.

When the granules are swollen and the crystallites are melted, the amylose molecules are freed to migrate into the surrounding aqueous matrix. In a limited-water system such as bread, only a small percentage of the starch is rendered soluble. In a fresh-baked product such as bread or cake, the starch granules are swollen, some of the amylose has migrated into the aqueous phase, and more of the amylose is at the granule surface, as are portions of some of the amylopectin molecules. The resulting granules have the appearance of hairy billiard balls. Several other factors influence the process, including the starch-water ratio, the presence of solutes such as sugar and salts, other components that compete with starch for water, and lipids that complex with the starch.

The temperature at which starch crystallites melt depends upon the amount of water available. At least two parts water to one part starch (by weight) are needed for all the starch to gelatinize in the normal temperature range (about 60–70°C for wheat starch). Bakery products all contain a water-starch ratio lower than this, so the temperature necessary to complete gelatinization is higher. For example, in a cake baked with 80% of the normal water content, the granules reach the first swelling stage at 96°C and never fully gelatinize. In a cake containing 120% of the normal water level, the granules swell at 88°C, and the cake is fully set before the end of the bake cycle. At the other extreme is a cookie dough that contains only 22% water. After baking, almost none of the starch granules are gelatinized. DSC studies of gelatinization at varying water-starch ratios indicate that when

---

Differential scanning calorimetry (DSC)—A method for measuring energy uptake as a sample is heated. When a phase change occurs (e.g., melting or freezing), the plot shows the temperature at which the change occurred and the amount of heat energy involved.

Birefringence—Ability of crystalline materials to rotate polarized light.

Gelatinization—Collapse (disruption) of molecular orders within the starch granule manifested by irreversible changes in properties such as granular swelling, native crystalline melting, loss of birefringence, and starch solubilization.

Gelatinization temperature—A narrow temperature range at which starch granules begin to swell, lose crystallinity, and viscosify the cooking medium. Starches from different sources have different characteristic gelatinization temperatures.

adequate water is present, the melting of starch crystallites is assisted by the solvent; i.e., they are hydrated as they melt. This occurs at what is usually called the *gelatinization temperature* (about 65°C for wheat starch). With limiting amounts of water, some portion of the starch crystallites melts and hydrates near 65°C, and the remaining portion melts during an anhydrous process that occurs over a temperature range of up to 115°C.

Sugar and salt also raise gelatinization temperatures. In amylograph tests, the addition of 2% sodium chloride raised the peak gelatinization temperature from 82 to 91°C. The elevation of gelatinization temperature by sugar depends in part upon the particular sugar used. In a 50% sucrose solution, the gelatinization temperature of wheat starch is raised by 26 degrees C, while in 50% glucose it increases by 20 degrees. This effect is particularly important in cake baking, in which coordination between the timing of starch granule swelling and the release of leavening gas, both of which are governed by internal batter temperature, is crucial for obtaining a satisfactory cake.

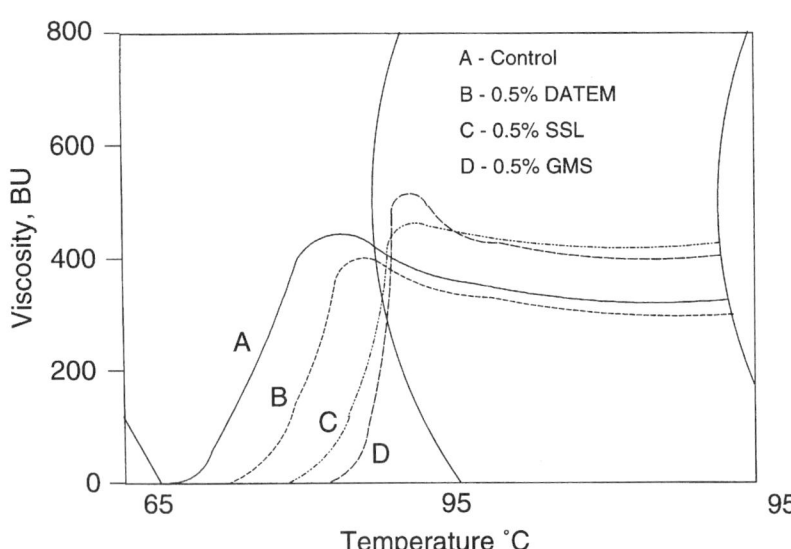

Fig. 4-1. Amylograph curves for wheat starch in the presence of diacetyl tartrate ester of monoglyceride (DATEM), sodium stearoyl lactylate (SSL), and glycerol monostearate (GMS).

The presence of other hydratable materials in a dough or batter decreases the amount of free water available for starch gelatinization. Studies on mixtures of gluten and starch showed that the gluten bound approximately 75% of its weight in water, and in a marginal system (0.91 g of water per gram of starch), this decrease in free water decreased the proportion of starch that gelatinized at the normal temperature (about 61°C) and increased the fraction that melted at a higher temperature. No doubt this also occurs when gums are added to cake batters for viscosity control.

Emulsifiers modify the gelatinization behavior of starch. Figure 4-1 shows the changes in *amylograph* gelatinization curves for wheat starch caused by the inclusion of various emulsifiers at a level of 0.5%. Of the three emulsifiers shown, diacetyl tartrate ester of monoglyceride (DATEM) is the least interactive, raising the swelling temperature by about 5 degrees C but not changing the viscosity of the gelatinized starch. Glycerol monostearate (GMS) has the most effect, raising swelling temperature by about 18 degrees C and increasing the paste viscosity. SSL has less effect on the inhibition of swelling, but it increases paste viscosity to about the same level as that caused by GMS.

Amylograph—An instrument used to study starch gelatinization. A slurry is heated from room temperature to 95°C at a set rate, held for a period of time, and then cooled at a set rate. The viscosity of the slurry is recorded as a function of time (hence, of temperature).

SSL and GMS reduce the level of solubilization of starch granules upon heating in excess water. At 85°C, 10% of the starch from a control sample is solubilized, but in the presence of 2% SSL or GMS, only 1.7% of the starch is solubilized. Monoglyceride produces this effect at higher temperatures, but at 95 and 120°C, starch solubility is the same in the presence of SSL as in the control. Measurements of swelling follow the same pattern. X-ray diffraction patterns of the insoluble portion of wheat starch heated in the presence of excess water and SSL or GMS show marked amylose-emulsion complex formation. The interaction can take place at the surface of the granules, and the starch-emulsion complex apparently serves to stabilize the granule, retarding water penetration and swelling as the temperature is raised. This occurs primarily with amylose. The inclusion of monoglyceride in a bread formula decreases soluble amylose in the crumb by about one-half but does not significantly change the amount of amylopectin that is solubilized.

STARCH RETROGRADATION

Amylose dissolved in water forms insoluble crystals, usually within hours. By contrast, amylopectin crystallizes much more slowly, taking days to do so. The difference in rate is probably caused by the relative differences in molecular mobility. The low molecular weight amylose is free to take different conformations and move around in solution, thus achieving molecular alignment and crystallization more readily than the high molecular weight amylopectin, which is sterically more hindered by branching. This reassociation is called *retrogradation* (1):

> Starch retrogradation is a process which occurs when starch chains begin to reassociate in an ordered structure. In its initial phases, two or more starch chains may form a simple juncture point which then may develop into more extensively ordered regions. Ultimately, under favorable conditions, a crystalline order appears.

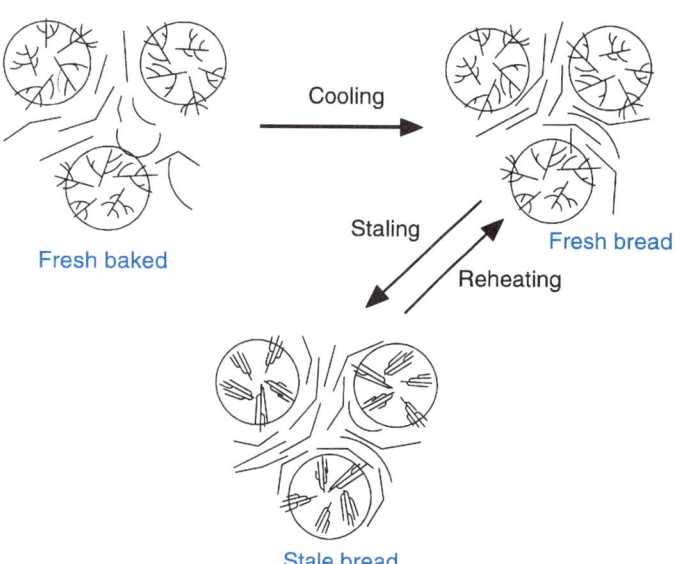

Fig. 4-2. Changes in starch molecules in bread during retrogradation (staling). In bread immediately out of the oven, both amylose and amylopectin are gelatinized and randomly oriented. During cooling, the amylose molecules align and crystallize. During storage, amylopectin reforms crystallites. (Reprinted from Stauffer, C. E., 1996, *Fats & Oils*, American Association of Cereal Chemists, St. Paul, MN)

The picture that best fits our present understanding of what occurs in baked products is shown in Figure 4-2. When bread or cake leaves the oven, the starch granules are gelatinized and a portion of the amylose is in the surrounding matrix. As the product cools, a significant portion (one-third or more) of the amylose crystallizes. During

Retrogradation—Recrystallization of gelatinized starch.

storage, the remainder of the amylose crystallizes, although this process appears to have relatively little to do with the increase in crumb firmness. The amylopectin forms crystallites during the several days of storage, causing the granules to become firmer (less compressible) and contributing markedly to increased crumb firmness.

The system is analogous to a sprayed concrete mesh structure in which the solid aggregate (gravel) contributes greatly to the strength and incompressibility of the whole system. If the gravel is omitted, the mesh is weaker and can be crushed more easily. The point of present interest is the change in slab compressibility as the "aggregate" changes from a soft gel to a firm, crystalline form.

Two physical phenomena are causally linked in this example: degree of crystallinity and elastic modulus (firmness). Quantitative X-ray analysis, *differential thermal analysis*, and the related DSC measurements provide information about crystallinity, while the various compression methods mentioned earlier measure elastic modulus. Data from several studies on bread and gelatinized starch gels show similar rate constants for the respective changes; the temporal correlation of crystallinity (indicated by thermal analysis) and firmness is good. On the other hand, a study that measured increases in firmness and crystallinity

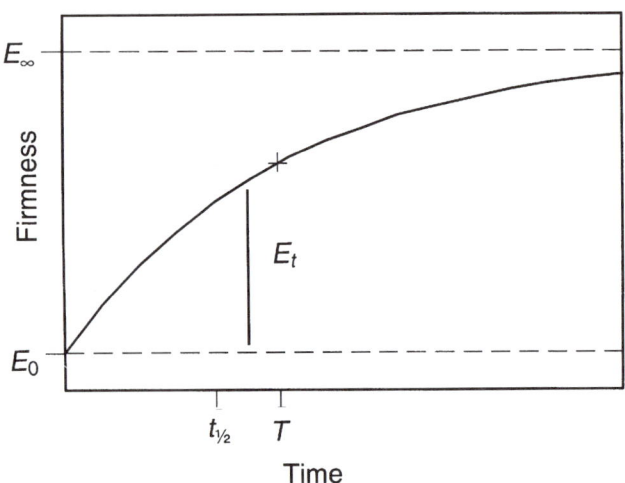

Fig. 4-3. Simple first-order rate process for the increase in firmness of bread crumb with time. E_0 is firmness at zero time, E_∞ is firmness at infinite time, and E_t is the measurement made at time t minus the zero time value. T is the reciprocal of the rate constant, and $t_{1/2}$ is the time for the value of $E_\infty - E_t$ to be reduced by half.

Differential thermal analysis—Method for measuring energy uptake as a sample is heated; similar to differential scanning calorimetry.

Box 4-1. Avrami Equation

Theoretical analyses of staling measurements are often done by using the modified Avrami equation to determine the level of crystallization in a restricted (solid) matrix. It considers that the fraction of crystallization that remains to take place, q, is an exponential function of some power of time, t (Fig. 4-3). That is

$$q = (E_\infty - E_t)/(E_\infty - E_0) = \exp(-kt^n)$$

in which E_0 is the initial value (of firmness, heat absorption, or other measure), E_∞ is the final or limiting value, and k is the rate constant.

While the ordinate in Figure 4-3 is labeled firmness, it can be any appropriate measure such as the index of crystallinity by X ray, heat absorption by DSC, or elastic modulus. For bread crumb and starch gels, a convenient unit of time t is days, usually resulting in k values of 0.1–1.

(continued on next page)

Box 4-1. *(continued)*

The exponent of time, n, is called the Avrami exponent and is related to the mode of crystallization. When the melted material contains no crystal nuclei, nucleation occurs sporadically because of random molecular motion. This happens in solutions that can be supercooled before some random event initiates crystallization. Instantaneous nucleation occurs when microscopic nuclei still exist in the melt and is often found in solutions that have been crystallized and then reheated. An example familiar to many bakers is high-fructose corn syrup. As it comes from the supplier, it remains clear even if it is cooled to below the nominal crystallization temperature for the glucose present, but if crystallization occurs, the syrup must be heated to a relatively high temperature and held for a long time to redissolve all the microcrystals of dextrose. The theoretical value of the Avrami exponent is the sum of two numbers: the first value equals the number of dimensions in which crystal growth occurs (1, linear; 2, circular; or 3, spherical) and the second value is 1 when instantaneous nucleation occurs and 0 when sporadic, random nucleation occurs.

An Avrami analysis of data requires a "double log" plot:

$$\ln[\ln(1/q)] = \ln(k) + n \ln(t)$$

The logarithm of the logarithm of the reciprocal of the fraction of change is plotted versus the logarithm of time. The intercept is the logarithm of the rate constant, and the slope of the plot equals n. In most studies of bread firming and starch gel retrogradation, the value found for n is 0.85–1. This seems correct for starch recrystallization. One would not expect all the intermolecular alignments (crystal nuclei) in starch to be totally disrupted, particularly under the condition of limited water that occurs during baking. The growth of the crystal (i.e., the addition of portions of the starch chains to the structure) would be expected to be one dimensional because of the linearity of the chains.

It is not clear from the basic theory what a value of $n < 1$ means in a physical sense. When a set of data is constructed by using the sum of two first-order reactions with rate constants that differ by a factor of 10 and the data are plotted according to the double log plot, a slope of <1 is obtained. Starch gels involve two different polymeric structures (amylose and amylopectin), and in bread, the complications are even greater. It appears that the complexity of the system may lead to an incorrect estimation of n. Overall, it is probably justifiable to assume a value of 1 for n in studies on bread and cake staling and starch gel retrogradation.

When $n = 1$, the Avrami equation is a first-order decay curve. This curve is completely described by the three terms E_0, E_∞, and k. Determining the parameters of a first-order reaction curve from experimental data can be easily done on a personal computer with one of the many simple programs available for this purpose. Frequently a value called the time constant, T, is reported ($T = 1/k$). At the point T (Fig. 4-3), 63.2% of the reaction has occurred. Another reference point often used is the half-life, $t_{1/2}$, which equals $\ln(2)/k$. At one half-life, the reaction is 50% complete, and $T = 1.44 \times t_{1/2}$.

Specific volume—In baking research, the weight of the cooled loaf divided by its volume.

(by X-ray analysis) on bread stored at 4 and 21°C found that the rate constant for crystallinity increase was about twice that of the rate for increase in firmness. It was also found that if the enzyme α-amylase is included in the dough, the relationship does not seem to hold. We may tentatively hypothesize a cause-and-effect relationship between starch crystallization and an increase in elastic modulus, realizing that the question still is not unequivocally settled. A more extensive discussion of this question has been published (2).

Starch retrogradation is influenced by a number of factors. Three of special interest to bakers are temperature, *specific volume*, and moisture content of the bread.

Staling has a negative temperature coefficient; that is, it accelerates as temperature decreases. This has been demonstrated many times. An Arrhenius plot of rate constant for staling (compression measurements) versus 1/temperature (°K) is shown in Figure 4-4. The plot is linear between 10 and 66°C, indicating that the rate of firming is a straightforward physical chemical phenomenon (the point on the right was obtained at –1°C). The straight line is fitted to the circled points, which represent bread made by the *Chorleywood bread process*. The triangles are points for bread made by a straight dough process with fermentation. This is objective proof of the general observation that bread stales faster when refrigerated.

Specific volume (g/cm³) does not affect the rate constant for firming, but as specific volume increases, the total change in firmness ($E_\infty - E_0$) decreases. The dependence of $E_\infty - E_0$ on specific volume is also process dependent. Chorleywood process bread shows a somewhat lower limiting firmness than fermented straight dough bread when the two are compared at the same specific volume. In addition, bread with a higher specific volume has a lower initial firmness, so at all stages it is softer than a more compact loaf.

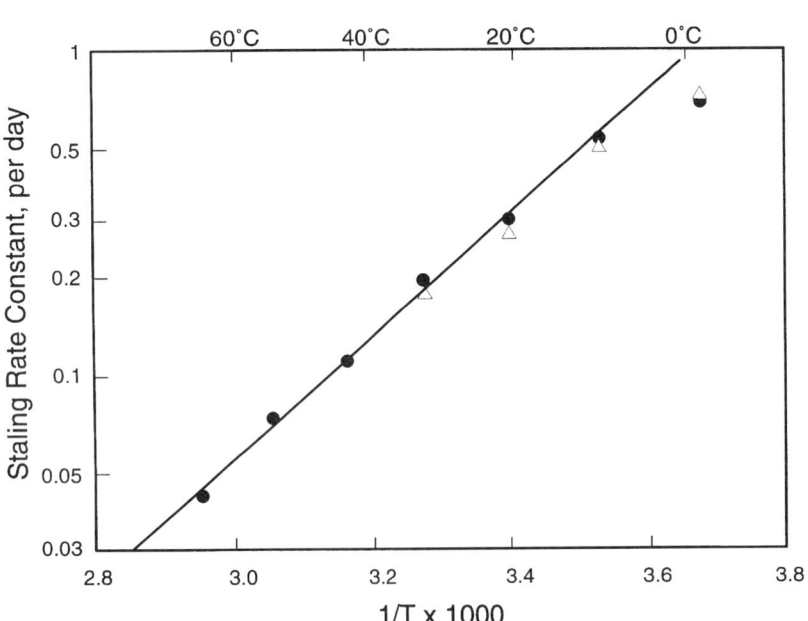

Fig. 4-4. Arrhenius plot of the effect of temperature on the first-order rate constant for the firming of bread crumb.

Chorleywood bread process—A rapid bread-making process, developed at the Chorleywood Laboratories in England, in which dough is mixed in a high-intensity mixer for a short period of time.

Starch retrogradation slows when the moisture content of the starch gel is high. To relate this to bread, remember that there is a moisture gradient in bread. In one study, the rate of retrogradation of crumb from the center of a loaf and from the region near the crust was determined by DSC. The center crumb had a moisture content of about 43% (45% initially and 42% after five days of storage at 21°C), and the crumb near the crust had a moisture content of approximately 32% (33% at day 1 and 31% at day 5). The firming of crumb from the center had a rate constant, k, of 0.22 per day, while that of crumb near the crust was 0.38 per day. The total amount of change was the same in both cases, 1.8 J/g. This confirms the general observation that moist bread stales less rapidly.

BREAD STALING

Several factors that are thought to influence bread staling rate are considered here briefly before the main discussion of the effect of surfactants on starch retrogradation and staling. These factors—moisture content, protein content, and processing variables—are not negligible from a practical standpoint.

Increasing the moisture content of bread increases its shelf life (as judged by the squeeze test). An obvious example is reduced-calorie bread, which contains 45–50% moisture versus the 38% level in standard U.S. white pan bread. The initial firmness of the high-moisture bread is slightly lower, and the compression curve remains lower throughout the normal shelf life. No data have been published that indicate whether there is a difference in the rate constant for firming and whether the total amount of firmness increase differs for regular bread and high-moisture reduced-calorie bread. As a rule, decreasing the finished moisture content of bread by 2% (e.g., by increasing bake time) shortens the shelf life by one day.

The major components of baked bread that bind moisture are heat-denatured gluten and gelatinized starch. Moisture migration within the bread crumb during cooling and storage has been studied, but results are contradictory. One group of researchers found that moisture migrates, if at all, from the retrograding starch to gluten, while another group, using a different approach, found migration to be in the opposite direction. In either case, the change in moisture of the starch portion is small, approximately 2%. Other indirect evidence for the role of moisture migration has been published, but the extent of the projected effect is minimal compared with the overall increase in crumb firmness.

It is known that increasing the protein content of bread tends to result in a softer loaf, which may be caused by simple dilution of the starch. Of course, increasing protein content also usually increases specific volume with attendant increased softness. An alternate hypothesis is that more protein binds more water, decreasing the moisture level in the starch phase and decreasing the rate or extent of starch crystallinity. However, this theory is not consistent with the known effect of moisture content on starch retrogradation. The role played by gluten protein in bread firming is not yet clear.

Processing parameters can markedly influence the shelf life (staling rate) of bread. Both overmixing and undermixing shorten shelf life, and long fermentation times extend shelf life. Slow baking shortens shelf life (moisture content is lower), while a fast bake lengthens shelf life (moisture content is higher). All of these factors are understood by industrial bakers, and adjustments in the baking process to optimize economic factors and production efficiency also tend to maximize shelf life.

EMULSIFIER-STARCH COMPLEXATION

Although starch and gluten proteins are usually thought of as hydrophilic molecules, readily wetted by water and becoming water soluble as the molecular size is reduced, they are in fact amphiphilic. Starch molecules are polymers composed of α-D-glucopyranosidyl residues joined primarily by 1,4 *acetal linkages* (1,6 linkages occur at the branch points in amylopectin). *Glucopyranoside* is a six-membered ring. It is not flat but is puckered in what is called the "chair" configuration (Fig. 4-5a). The bond angles from each carbon are such that the hydrophilic hydroxyl groups project outward to the side of the plane of the ring, while the hydrogen atoms project either above or below this plane. The perimeter of the ring is hydrophilic, and the two faces are hydrophobic. The bond angle of the α-1,4 acetal linkage is such that the starch chain coils to form a *helix*, with about six residues per turn (Fig. 4-5b). It is difficult to draw the details of this helix, but molecular models show that the plane of the residue ring lies parallel to the wall of this helix and that the hydrogen atoms on carbons 3 and 5 (circled in Fig. 4-5a) project into the interior of the helix. The result is a hollow cylinder that has a hydrophilic outer surface and a hydrophobic inner surface. This inner space is about 4.5 Å in diameter, and straight-chain alkyl molecules such as stearic acid can fit into it (Fig. 4-5c), as can other molecules such as iodine. The blue color of an iodine-starch complex demonstrates that iodine is in a nonpolar environment; iodine dissolved in chloroform is blue, but in water it is brown. The complex of amylose with *n*-butanol crystallizes much more readily than amylose alone, and this behavior has been used to separate amylose from amylopectin.

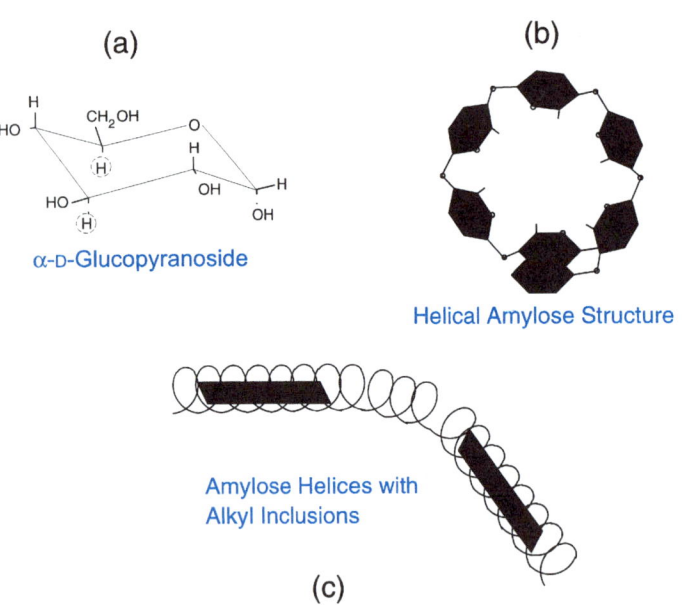

Fig. 4-5. Structures of α-D-glucopyranoside (a), the amylose helix (b), and the inclusion complex of amylose with straight-chain lipids (c).

Acetal linkages—Bonds between sugar residues in polymers, linking the carbonyl group of one residue to a hydroxyl group on the other sugar.

Glucopyranoside—Glucose in its usual molecular form of a six-membered ring.

Helix—Three-dimensional arrangement of many biological polymers, including starch. It is analogous to a coil spring.

The *n*-alkyl portion of emulsifiers such as GMS forms a complex with helical regions of starch. This interaction has been measured

Table 4-1. Starch-Lipid Complex Formation

	Excess Emulsifier with Amylose[a] (mg/g complexed)	Excess Emulsifier with Amylopectin[a] (mg/g complexed)	Excess Amylose (ACI)[b]
1-Monocaprin	...	...	63
1-Monolaurin	...	...	95
1-Monomyristin	23.4	...	100
1-Monopalmitin	37.4	5	72
1-Monostearin	33.9	8.3	87
1-Monoarachidin	...	11	...
1-Monobehenin	...	...	16
1-Monoolein	24.6	...	9
1-Monoelaidin	...	...	75
1-Monolinolein	12.2	...	0

[a] Data from (3).
[b] Data from (4) and (5). ACI = amylose complexing index

quantitatively but with two different approaches: excess emulsifier and excess starch (Table 4-1). In the first approach, 1 g of amylose or amylopectin was heated with 100 mg of monoglyceride. The mixture was cooled, and the amount of uncomplexed monoglyceride was determined. The results were given as milligrams of lipid bound per gram of starch. In the second approach, 100 mg of amylose was heated with 5 mg of emulsifier. Upon cooling, the amylose-emulsifier complex precipitated, and the amount of dissolved, uncomplexed amylose was quantitated by its affinity for iodine. The results were expressed as an amylose complexing index (ACI), which is the percentage of the amylose complexed by 5 mg of emulsifier.

The results can be summarized as follows:

1. Saturated fatty acids (12–20 carbons long) are the best complex formers.
2. *cis* Unsaturated C18 fatty acids are very poor complex formers.
3. *trans* Unsaturated C18 fatty acids are good complex formers.

There is a definite correlation between the ability of a monoglyceride to form a complex with starch and its ability to retard the rate of staling (i.e., the increase in crumb firmness). This has been confirmed in several studies, which showed, for instance, that 1% monopalmitin or monostearin increased shelf life by as much as two days, whereas bread made with 1% monoolein or monolinolein staled at the same rate as the control. Also, the monoglyceride should be present in the dough as the lamellar mesophase. This may account for the *synergistic* functionality of some mixed systems (e.g., GMS plus SSL or GMS plus Polysorbate 60).

The mechanism by which emulsifiers slow crumb firming (staling) may be summarized as follows:

Synergistic—Pertaining to a combination of two materials that displays more functionality than would be expected by simply summing the individual functionalities of the materials.

1. Crumb firmness increases during storage primarily (though not exclusively) because of retrogradation of amylopectin molecules.
2. Emulsifiers form complexes with gelatinized amylopectin, hindering its ability to recrystallize.
3. Emulsifiers also form complexes with solubilized amylose and may hinder its ability to contribute to formation of a solid intergranular starch matrix.

The model of staling resulting from starch retrogradation was developed by T. Schoch during the 1950s, and the role of emulsifiers in retarding staling was explicated during the 1960s and 1970s, mainly by W. Knightly and N. Krog. Subsequent work has indicated some particular subareas of this phenomenon that are still puzzling, but the main concept seems to be valid.

As discussed in Chapter 3, monoglyceride is added to bread and rolls to decrease the rate of staling. Shelf life extension increases up to a level of about 0.75–1% monoglyceride (flour basis); more emulsifier produces relatively little effect. In using the various forms of monoglyceride, the baker should keep in mind that they contain different levels of the active material. The approximate relationship is 4 oz. of powdered distilled monoglyceride = 8 oz. of plastic monoglyceride and diglyceride = 16 oz. of hydrated monoglyceride.

It should also be remembered that both 1- and 2-monoglycerides are effective antistaling agents. Some product analyses report total monoglyceride, while others report only 1-monoglyceride (sometimes called α-monoglyceride), which represents about 92% of the total monoglyceride.

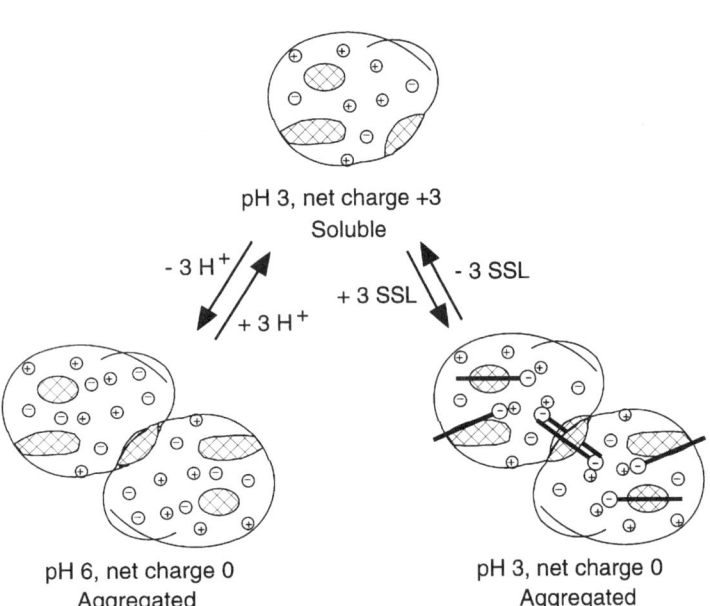

Fig. 4-6. Influence of pH and anionic surfactant molecules on gluten protein. The crosshatched areas depict nonpolar patches on the protein surface. SSL = sodium stearoyl lactylate.

Hydrophobic patches—Areas on the surface of a protein molecule, in contact with surrounding water, that are lipophilic in nature.

Dough Strengtheners

Some of the amino acid side chains in proteins are hydrophobic, generally buried in the interior of the folded protein molecule but exposed if the protein is unfolded. Frequently, these hydrophobic regions are partially exposed, even in the native folded protein, and are referred to as *hydrophobic patches* on the protein surface (Fig. 4-6). The lipophilic parts of surfactants interact with these hydrophobic regions, sometimes contributing to unfolding (denaturation) of the protein and further binding of surfactant. Gluten protein contains about 40% hydrophobic amino acids, and it interacts strongly with lipid-type materials. In a mixed dough, more than half the native lipid plus any added surfactant is bound to the protein, while the rest is uncomplexed and freely extractable. The surfactant SSL binds to gluten proteins as do *fluorescent probes* for hydrophobic environments. It can safely be said that gluten proteins have rather large hydrophobic regions on their surfaces, and some of the properties of gluten may be explained on this basis.

Two of these gluten characteristics are depicted in Figure 4-6. When acid is added to a flour-water dough, some of the protein solubilizes. This is most likely a charge effect. At a dough pH of about 6, the gluten proteins carry a low charge density resulting from an approximate equality between the cationic amino acids (lysine, arginine, and histidine) and the anionic amino acids (glutamic and aspartic acids). As the pH is lowered, many of the anionic carboxylates are protonated (i.e., become un-ionized), and the protein molecule takes on an overall positive net charge. At pH 6, the hydrophobic patches can interact, and the protein molecules aggregate via hydrophobic interaction. At pH 3, the net positive charges cause the molecules to repel each other, and solubilization occurs. This situation has many similarities to that of emulsified oil droplets stabilized by an ionic surfactant, where the surface charge prevents droplet contact and coalescence. Likewise, salt represses the electrostatic repulsion, and protein aggregation is favored. Most dough strengtheners are anionic surfactants, and when the lipophilic tail of the surfactant binds to the protein hydrophobic patches, it incorporates this negative charge into the complex, moving the overall charge closer to zero and promoting aggregation in the dough (Fig. 4-6). Salt and SSL have similar effects on mixograph curves, and it might be said that salt suppresses the electrostatic repulsion while SSL neutralizes it. The final effect in both cases is the same, i.e., hydrophobic aggregation of the gluten protein and an increase in dough strength.

An excess amount of surfactant can solubilize proteins, and additional adsorption to the protein generates an excess net charge, even if the protein net charge is near zero. Sodium stearate (4 mg per 10 mg of glutenin) renders glutenin proteins completely soluble in distilled water. Cationic surfactants also solubilize glutenin, but nonionic or amphoteric surfactants (e.g., Tweens, Span, and lecithin) have little or no solubilizing effect.

The usual test of a dough strengthener is a bake test, in which the loaf volume increment resulting from the inclusion of the surfactant being tested is determined. This change in loaf volume appears to have more to do with the interaction of the surfactant with the starch granule than with the effect upon gluten protein. The final loaf volume of a dough with 3% shortening is greater than that of a control loaf containing no shortening, because the loaf containing shortening expands for a longer time during the bake cycle. The presence of shortening delays the swelling of starch granules (and perhaps the denaturation of the gluten protein), and this delay translates into a larger loaf volume (i.e., larger *ovenspring*). The addition of a surfactant such as SSL or DATEM to dough also produces this delay in the setting mechanism

Fluorescent probes—Small molecules that fluoresce in a nonpolar medium but not in a polar medium such as water. When these probes are mixed with proteins, for example, the appearance of fluorescence implies that a probe has bound to a hydrophobic region of the protein.

Ovenspring—Increase in the volume of a loaf of bread during baking; i.e., the final loaf volume minus the volume of the dough at the end of the proof.

and thus increases the loaf volume. Numerous studies with a variety of emulsifiers (e.g., sucrose mono- and diesters, calcium stearoyl lactylate [CSL], DATEM, and GMS) report similar results. It should be noted that the surfactants are about 10 times more effective than triglycerides in producing loaf volume enhancement by this mechanism; 0.2–0.4% surfactant gives the same effect as 3% shortening.

A more meaningful test of the dough-strengthening capabilities of a surfactant involves subjecting the proofed loaf to mechanical abuse before putting it into the oven. The pan containing the proofed dough is placed on blocks of wood 3.75 in. high. The blocks are pulled from under the pan, allowing it to fall to the counter top. This procedure is conducted two more times, and then the doughs are baked, depanned, and cooled, and the volume is measured. In a study in which this technique was employed and the use level of all strengtheners was 0.25%, CSL produced the best volume improvement (approximately 260 cm^3 in a 1-lb loaf), ethoxylated monoglyceride was the second most effective, Polysorbate 60 and SSL were roughly equivalent (about 210 cm^3), and succinyl monoglyceride showed the least improvement (about 125 cm^3) compared with the control loaf. This test is designed to simulate the rough handling, such as sudden starts and stops on the conveyor line or being struck by a loading or unloading bar, that proofed doughs often undergo in a commercial bakery. A more reproducible test is to allow the pan to slide down an inclined plane (a short section of the conveyor) and hit a stop bar at the bottom. Although the literature is replete with data on delaying starch swelling and volume improvements such as those discussed above, when a dough strengthener is evaluated for commercial use, some sort of abuse test of the proofed dough is the best way to determine dough-strengthening capabilities of a surfactant.

The basic mechanism by which dough strengtheners work appears to be enhancement of gluten protein aggregation, either by charge neutralization (anionic surfactants) (Fig. 4-6) or by some sort of hydrogen bonding (ethoxylated surfactants). The surfactants are bound rather strongly to the native gluten protein, but they are not bound to the heat-denatured protein after baking. CSL has little effect on the mixing requirements of dough but increases its tolerance to overmixing, while the anionic surfactant sodium dodecyl sulfate markedly increases the mixing tolerance of a dough in the mixograph.

It might be that the fatty acids of dough strengtheners form a complex with gelatinized starch and thus contribute to antistaling activity. SSL and DATEM are reported to have ACI values of 72 and 49, respectively. Reports from baking and storage tests in which these surfactants were used indicate that they do indeed retard crumb firming compared with the control. However, in most such studies, loaf specific volume was not controlled, which muddles interpretation of the results.

Studies with lidded bake pans (which keep loaf specific volume constant) indicate that DATEM does preserve crumb softness. The effect of combining DATEM with GMS appears to be roughly multiplicative: at day 7, the bread containing 1% DATEM had 53% of the firmness of the control; the loaves with 1% GMS had 40% of the firmness of the control; and the bread containing 1% GMS plus 1% DATEM registered 22% of the firmness of the control. This relationship implies that the two emulsifiers are exerting their antistaling effects through different mechanisms and are not competing for the same complexing sites in the gelatinized starch. The researchers propose that DATEM reduces crumb firmness by affecting crumb cell wall thickness and elasticity and that its effect on starch retrogradation is limited. In other words, the gluten protein in bread crumb plays some role in staling, and emulsifiers such as DATEM and SSL modify this contribution.

Aeration Agents

Air is entrained in most bakery products during the preparation stages. In baked items (excluding icings and the like), the degree of subdivision of the air bubbles determines the nature of the finished crumb. The air bubbles are nuclei for the leavening gases generated during baking. If the air is present as a few large bubbles, the finished product will have a coarse crumb, while if the air is divided into many small bubbles, the finished crumb will be smooth and fine grained. The presence of emulsifiers (either natural or added) aids in air dispersal and inhibits coalescence of the bubbles during the processing period before baking. This was first demonstrated for bread in 1940 and subsequently for other products such as cakes, doughnuts, and cookies.

Successful manufacture of a good-quality layer cake requires dispersal of air throughout the batter and retention of the bubbles until the starch has swollen and the cake structure is set. Traditionally, air has been entrained in the shortening phase during the first stage of making a cake. This method is not an efficient way to incorporate air, because it depends primarily upon the ability of the plastic shortening to trap air bubbles during creaming. The "superglycerinated" shortenings containing monoglycerides, developed in about 1930, enable the baker to subdivide the air bubbles, creating smaller bubbles that are more efficiently retained by the shortening phase and that give more uniform nucleation for leavening gases throughout the batter during baking. The final crumb grain, therefore, is closer and the overall volume is larger (Fig. 4-7). These

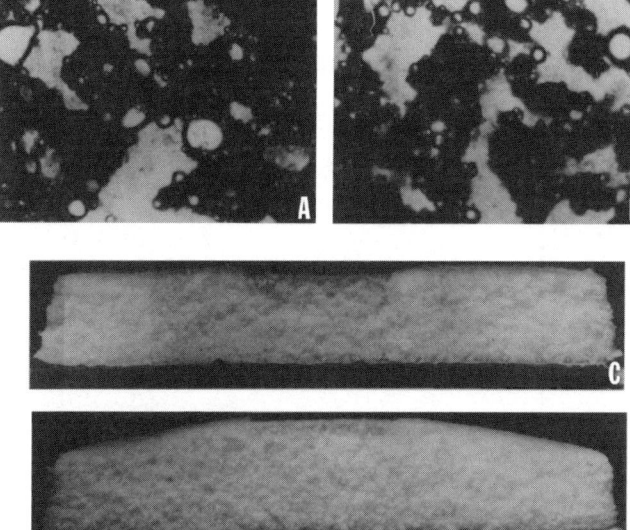

Fig. 4-7. High-ratio cake batters made by three-stage mixing. A, Plastic shortening, no emulsifier, batter specific gravity 0.85 g/ml. B, Plastic shortening containing 4.5% monoglyceride, batter specific gravity 0.81 g/ml. C and D, Cakes baked from A and B, respectively.

shortenings also enable the production of high-ratio cakes (i.e., cakes containing more sugar than flour).

This traditional approach involves at least three stages in batter preparation: first, shortening and sugar are creamed vigorously to incorporate air; second, eggs are blended into this whipped material; and third, the flour, milk (or water), flavors, and other ingredients are added and blended to make the final batter. The development of dry cake mixes for home and industrial use dictated a different approach to air incorporation. Using a liquid vegetable oil in place of plastic shortening results in a moister cake and a longer shelf life. However, a liquid oil is not well suited to entrapping air bubbles. It was found that the addition of certain emulsifiers to the oil before it was incorporated into the cake mix produced high-ratio cakes with good volume, fine grain, and excellent keeping qualities. These cakes can be mixed in one stage. All the ingredients are placed in a bowl, the liquid is added, and the batter is mixed at low speed (to blend ingredients) and then at high speed to incorporate air.

The most useful emulsifiers for this purpose are the α-tending emulsifiers, because they tend to solidify in a stable α-crystalline form. Those used on a commercial basis today include acetyl monoglycerides (AcMG), lactyl monoglycerides (LacMG), and propylene glycol monoester (PGME). A number of other emulsifiers have also been used, including Polysorbate 60, stearoyl lactylic acid, and sucrose esters, but most commercial cake emulsifiers now offered by suppliers are based upon PGME and/or AcMG, perhaps in combination with other emulsifiers that enhance their functionality.

When dissolved in the oil phase, α-tending emulsifiers lower interfacial tension, but their effectiveness in cake batters is not the result of this property. Rather, at concentrations above a certain level, these emulsifiers form a solid film at the oil-water interface. This behavior is seen when one suspends a drop of water from a syringe tip in the oil, waits a few minutes for film formation, and then withdraws some of the water (Fig. 1-10). The film appears to be caused by crystallization of the emulsifier at the interface in the α form. There is a defined relationship between temperature and the minimum bulk concentration of the emulsifier that produces a film; the minimum concentration increases as temperature increases (Fig. 4-8). The addition of a second surfactant may enhance film formation. A mixture of propylene glycol monostearate (PGMS) and stearic acid (80:20) forms a stronger film than pure PGMS at the same weight concentration.

In a one-stage cake batter, air incorporation is primarily a function of foam stabilization by the protein contributed by flour, milk, and egg whites. As discussed earlier, the presence of oil inhibits this foam formation, and the solid film at the oil-water interface effectively encapsulates the oil during air incorporation, thus preventing destabilization.

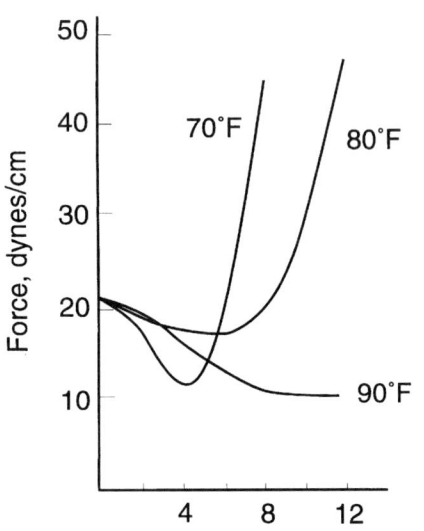

Fig. 4-8. Effect of temperature on film formation by 1-acetyl-3-monostearin at the oil-water interface. The sharp rise in the tensiometer reading indicates the presence of the solid film. CSO = cottonseed oil.

Simply lowering cake batter density by incorporating more air is not enough to ensure a good final volume. In a comparison of two sucrose esters in a sponge cake formula batter, densities were 0.65 and 0.55 for F-110 and F-160, respectively, but F-110 gave a larger finished volume than F-160 (see Table 3-1). Similarly, stearoyl lactylic acid decreases batter density markedly as its usage level is increased from 0 to 4% (flour basis), but the best finished cake volume is achieved at 1.5% emulsifier. These results merely confirm the earlier remark that not only is it necessary to incorporate air into a cake batter during mixing, but also that air bubbles must remain in the batter during baking, serving as nuclei for the gases released from the chemical leavening system.

Air incorporation is also important in obtaining a fine grain in cookies, cake doughnuts, and similar products. If a cookie is made with nonemulsified plastic shortening, for example, it will have a more open texture than the same product made with an emulsified shortening (containing 3% α-monoglyceride). The aeration of creme icings is enhanced by the presence of an emulsifier in the shortening used. However, while initial aeration increases as α-monoglyceride concentration is increased up to 4%, the best icing stability is observed at 2% emulsifier. At higher levels, the air escapes more readily from the icing during storage. The reason for this behavior is not known.

Whipping aids such as sodium lauryl sulfate (SLS) or triethyl citrate enhance foam formation when egg whites are whipped. The surfactant aids the unfolding of the protein, which is the actual foam stabilizer. A small amount of the surfactant is effective. SLS is added at a maximum level of 1% to dried egg whites for this functionality.

Troubleshooting

In bakery products, emulsifiers work with the fat phase, protein, and starch to produce a good final product. Often, a mixture of emulsifiers works better than a single emulsifier. For many problems, the solution may require adjustments not only to the emulsifier system but also to the other ingredients or the processing parameters. In emulsion systems, troubleshooting is still more an art than a science, because many of the interactions taking place between the ingredients and the processing parameters are specific to each food product. While the following table suggests adjustments to be made in emulsifier use, the reader can also consult the troubleshooting sections in the Fats & Oils handbook that deal with shortening-related problems.

BREAD		
Symptom	Cause	Changes to Make[a]
Sidewall weakness ("keyholing")	Inadequate strength of crumb structure and low elasticity	Add dough strengthener (e.g., SSL or DATEM) up to legal limit (0.5% flour basis for SSL and good manufacturing practice for DATEM).
Open grain	Inadequate gluten strength	Add dough strengthener (e.g., SSL or DATEM).
Low volume (sometimes noted when substituting oil for plastic shortening)	Inadequate gluten strength	Add dough strengthener (e.g., SSL or DATEM).
Short shelf life	Insufficient staling inhibitor	Use up to 0.5% actual α-monoglyceride (flour basis). Use easily dispersible form. (Excess could cause open cell defect.)
Sticky dough	Gluten not fully developed; process water too soft or too alkaline	Adjust water pH (e.g., with lactic acid or monocalcium phosphate).
YEAST-RAISED DOUGHNUTS		
Symptom	Cause	Changes to Make
Short shelf life	Inadequate fat absorption	Add 0.5% α-monoglyceride to base dough.
Collapse on cooling	Weak gluten structure	Add SSL or DATEM up to legal limits.
CAKES		
Symptom	Cause	Changes to Make
Tunnels in the cake	Inadequate emulsification	Use emulsified shortening (e.g., 3.5–5% α-monoglyceride). Add emulsifier with a high HLB (e.g., 0.25% Polysorbate 60) to the batter.
	Too hot an oven	Lower oven temperature.
	Excess leavening	Lower leavening level.
	Inadequate hydration and dispersion of ingredients	Allow enough time to make up batter.
Low volume; center dip	Inadequate emulsification; poor creaming of fat	Use an emulsified shortening (e.g., 3.5–5% α-monoglyceride).
	Poor foam stability (oil cakes)	Increase level of α-tending emulsifier (e.g., PGME or AcMG) at 12–14% of oil.
	Excess liquid	Decrease water level.

[a] AcMG = acetylated monoglyceride; DATEM = diacetyl tartrate ester of monoglyceride; HLB = hydrophilic/lipophilic balance; PGME = propylene glycol monoester; and SSL = sodium stearoyl lactylate.

CAKE DOUGHNUTS		
Symptom	**Cause**	**Changes to Make**
Excessive fat absorption	Excess emulsification	Reduce amount of emulsifier in the formula.
	Cold batter	Use warmer water during batter make-up.
	Batter viscosity too high	Increase amount of water in batter formula.
	Under mixed	Increase mixing time to fully hydrate ingredients.
	Fryer temperature too low	Raise fryer temperature to recommended level. Make sure heater can keep up with rate of batter deposition.
	Fat breakdown	If fryer is not in continuous use, lower fryer temperature between fries. If use is continuous, match doughnut production with fryer size to get complete fat turnover in 8 hr.
Low fat absorption	Insufficient emulsification	Add 0.5% α-monoglyceride or 1% lecithin to the formula.
	Hot batter	Use cold water (or ice) to reduce batter temperature.
	Batter viscosity too low	Reduce water in batter formula.
	Over mixed	Decrease mixing time.
	Too much floor time before depositing	Balance batch size and mixing schedule with rate of production in fryer.
	Fryer temperature too high	Lower fryer temperature to recommended level.
	New (untempered) fat	When refilling fryer (after cleaning), use old (not degraded) fat equal to one-fourth of fryer capacity. If old fat is not available, add 1.5 oz free fatty acids per 100 lb new fat.

CREME ICINGS		
Symptom	**Cause**	**Changes to Make**
Low aeration (high specific gravity)	Inadequate emulsification	Use shortening with 2–3% α-monoglyceride. Use an emulsifier with a high HLB (e.g., up to 0.46% Polysorbate 60 based on total icing).
	Fat too soft to entrap air	Use shortening with a high solid fat index profile.
	Inadequate creaming	Increase first stage mixing time of fat plus sugar.
Loss of aeration during storage	Unbalanced formula	Rebalance HLB of emulsifier system. Add hydrocolloid stabilizer.
	Excess monoglyceride	Use shortening with a maximum of 3% monoglyceride.
	Over or under whipping	Adjust whipping time.

References

1. Atwell, W. A., Hood, L. F., Lineback, D. R., Varriano-Marston, E., and Zobel, H. F. 1988. Cereal Foods World 33:306-311.
2. Hebeda, R. E., and Zobel, H. F., Eds. 1996. *Baked Goods Freshness.* Chapter 1, The Staling Mechanism, and Chapter 2, Surfactants. Marcel Dekker, New York.
3. Lagendijk, J., and Pennings, H. J. 1970. Cereal Sci. Today 15:354-356, 365.
4. Krog, N. 1971. Starch/Stärke 23:206-210.
5. Riisom, T., Krog, N., and Eriksen, J. 1984. J. Cereal Sci. 2:105-118.

Supplemental Reading

Stauffer, C. E. 1990. *Functional Additives for Bakery Foods.* Chapter 3, Emulsifiers and Dough Strengtheners. Van Nostrand Reinhold, New York.

Thomas, D. J., and Atwell, W. A. 1999. *Starches.* American Association of Cereal Chemists, St. Paul, MN.

CHAPTER 5

Dairy and Nondairy Products

Milk is a protein-stabilized oil/water (O/W) emulsion. The presence of protein at the interface complicates the interpretation of experimental results and the understanding (at the molecular level) of the events that occur when milk-derived and milk-related products are processed. These complications increase when emulsifiers such as those discussed in Chapter 3 are added to the system. How proteins act at air-water and oil-water interfaces is also discussed in Chapter 3.

Milk

Milk proteins are divided into two groups: *casein* and the *whey* proteins. Casein (the protein that precipitates to make cheese) is a complex containing four major proteins, α_{s1}-, α_{s2}-, β-, and κ-casein, in the rough proportion of 40:10:40:10. The molecular weights of the caseins range from 24,000 for β-casein to 121,000 for α-casein. In addition, the caseins aggregate to form micelles with molecular weights of approximately 1×10^9. Casein is often dissolved at alkaline pH and then neutralized and dried to create compounds such as sodium caseinate, which has an estimated molecular weight of 250,000. In whey, the main proteins are α-lactalbumin (approximate molecular weight, 17,000) and β-lactoglobulin (approximate molecular weight, 45,000); numerous others are present in minor amounts. The ratio of hydrophobic to hydrophilic amino acid residues varies greatly. As a rule, the larger the protein, the greater the proportion of hydrophobic residues. Given these disparities in size and relative hydrophobicity, one would predict that interfacial behavior might vary widely among different proteins, and this is indeed the case. The behavior of whey protein concentrate differs from that of sodium caseinate, which in turn differs from that of β-casein. When the numerous possible combinations of proteins at the interface are also considered, it is easy to understand why the interfacial chemistry of milk is so complex.

Milk fat droplets are synthesized in the endoplasmic reticulum of mammary epithelial cells. They are released into the glandular intracellular region after being surrounded by a plasma membrane, which is a highly complex, stratified structure consisting mainly of polar lipids and proteins. The two major proteins of the membrane are xanthine oxidase (molecular weight, 155,000) and a hydrophobic glycoprotein, butyrophilin (molecular weight, 67,000). The structure of

---

In This Chapter:

Milk

Butter and Margarine
 Butter
 Margarine

Whipped Cream and Nondairy Whipped Toppings
 Whipped Cream
 Nondairy Whipped Toppings

Ice Cream

Coffee Whiteners

Troubleshooting

Casein—The main protein component of milk, accounting for about 80% of the total proteins.

Whey—The liquid left after casein has been precipitated from milk. In addition to protein, it contains lactose (milk sugar) and ash (inorganic salts).

Milk fat—The natural fat found in milk consisting of a mixture of glycerides.

this membrane is different in the milk from various mammalian species. This discussion relates only to cow's milk.

The diameters of milk fat droplets in native milk range from 0.2 to 20 µm; 90% of the diameters are 1–8 µm, and the average diameter is about 3 µm. Since the droplets are less dense than the aqueous portion, they rise to the top of unagitated milk within a day and are known as "cream." The droplets are highly resistant to coalescence, and the cream can be easily redispersed by simply inverting the container a few times. Cooling milk enhances the creaming tendency by causing an interaction between the milk membrane proteins and some immunoglobulins present in the aqueous phase. Centrifugation of milk accelerates the separation of the fat phase and is done routinely. The fat content of unprocessed milk may vary from 2.5 to 4%, but for consumer sales, the fat content is standardized at specified levels (0, 0.5, 1, 2, or 3.5%). The cream is processed separately for other products.

After the fat content is standardized, the milk is homogenized by forcing it at high pressure through a valve homogenizer. The typical average fat droplet diameter decreases from 3 to <1 µm, and the surface area per milliliter of fat increases. The additional interfacial area requires stabilization, which is supplied by proteins from the aqueous phase. About 95% of this protein is casein, some in the form of the various individual casein species and some as deformed casein micelles. Whey proteins (lactalbumin and lactoglobulin) are incorporated into the interfacial protein membrane in varying proportions depending on homogenization conditions, especially temperature. Denaturation of whey protein leads to formation of casein-whey complexes, which adsorb more readily at the interface. The adsorbed protein layers are quite stable. Simple desorption of protein from an oil-water interface is quite slow (a period of days), and there is also some evidence that crosslinking occurs between the adsorbed proteins, which further increases stability.

Heat treatment of the milk promotes flocculation of the fat globules, presumably through interaction of heat-unstable whey proteins in the interfacial layers. Again, the fat droplets do not coalesce but clump together and are readily redispersed by mild agitation.

Butter and Margarine

BUTTER

Churning cream to make butter is essentially a *phase inversion*, i.e., transformation of an O/W emulsion into a water/oil (W/O) emulsion (Fig. 5-1). The basic process has been used for millennia, but it has been modified extensively in recent years to make it more amenable to large-scale industrial butter production.

During churning, the fat globules are forced together, some degree of membrane removal occurs, and the fat coalesces. This process continues until the fat forms a discrete clump with some entrained aque-

Phase inversion—Conversion of the continuous phase of an emulsion to the discontinuous phase and vice versa.

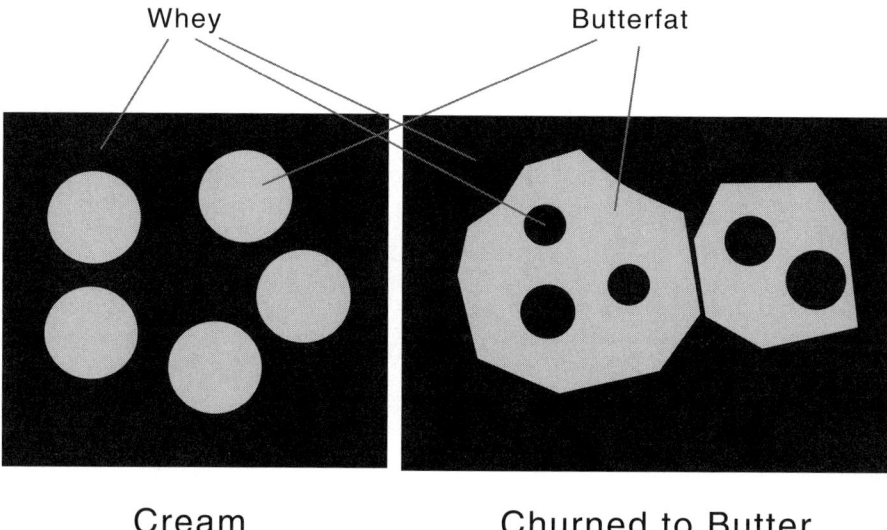

Fig. 5-1. Coalescence of milk fat globules during churning, resulting in an inversion from an oil/water emulsion to a water/oil emulsion.

ous phase. When all the fat has been transformed in this fashion, it is separated from the residual buttermilk; additives (e.g., salt and coloring) are added; and it is kneaded into a smooth, homogenous mass. In most countries, there are regulations about the composition of butter (e.g., a minimum of 80% milk fat).

The temperature during churning is an important processing parameter. Fat globules coalesce from the liquid phase, but the solid fat fraction of the milk fat stabilizes the clump and keeps the entrained aqueous droplets separate. The *solid fat content* of milk fat is lower during the winter than during the summer, and this difference must be accounted for in the churning operation. If the milk fat is too hard, the protein membrane does not desorb sufficiently to allow coalescence; if it is too soft, the aqueous phase and air (about 3–5% of the total mass by volume) are not incorporated. (The proper water and air content is important for the flavor and spreadability of the butter.)

Two things contribute to the stability of the water droplets: the fat protein membrane and crystals of high-melting-point triglycerides from the fat. When an emulsion inverts, the emulsifier tends to remain at the interface. As shown in Figure 5-1, the curvature of the membrane simply reverses and appears to encase the aqueous phase rather than the fat phase. The reason for desorption of the membrane is not fully understood, but it is probably the result of mechanical forces rather than interfacial energetics. As mentioned above, protein desorption from an interface is a slow process, but during churning, it occurs quickly. Thus, it seems likely that when two fat globules collide, the membrane is physically ruptured (made easier because the fat phase is semiliquid), allowing the globules to make contact and

Solid fat content—A measure of the amount of solid fat in a fat at various temperatures, determined by nuclear magnetic resonance.

Margarine—A substitute for butter, originally formulated to mimic butter as closely as possible, but using other fat sources in place of milk fat.

coalesce. The involvement of fat crystals is somewhat less easy to explain but is not an unknown phenomenon. Solid particles tend to stabilize an interface, probably by simply interfering with the contact and coalescence of the dispersed phase.

MARGARINE

Originally invented as an inexpensive substitute for butter, *margarine* today is purchased for its own characteristics. It is a W/O emulsion. The oil is generally some blend of partially hydrogenated vegetable fats, chosen to meet a specified solid fat content profile. The water phase varies greatly from reconstituted nonfat dry milk to water with some added flavors.

Standard margarine contains at least 80% fat; the rest is a skim milk, whey, or water solution containing salt and flavors. In preparing margarine, there are six steps:

1. blending base stocks plus other oil-soluble components;
2. mixing milk or water with salt and other water-soluble ingredients;
3. mixing the two phases to form a W/O emulsion;
4. chilling and plasticizing the emulsion;
5. forming the margarine into prints or placing it into plastic tubs; and
6. finish packaging and cold storage of the finished product.

In step 1, the oils to be used are weighed into a tank (at 5 degrees C above their melting points), and emulsifier (lecithin, mono- and diglyceride, or a combination), oil-soluble vitamins (A, D, and E) and colorants (β-carotene and annatto) are added. The oil is mixed, preferably under a nitrogen atmosphere to prevent oxidation and flavor deterioration.

When the oil-soluble ingredients are thoroughly mixed, the water phase is added. This can be whole milk, skim milk, reconstituted nonfat dry milk solids, or water. Salt and flavors such as diacetyl and starter distillate are included in the water phase. Antimicrobial agents (sodium benzoate and potassium sorbate) and/or heavy metal chelating agents (citric acid or EDTA) are added if needed and if the law allows it. The water phase is slowly added to the oil with agitation to form a W/O emulsion. Sometimes the emulsion is pasteurized at 73°C for 16 sec to ensure freedom from pathogenic bacteria.

The emulsion is quickly chilled in a swept-surface heat exchanger. The violent agitation and kneading in this unit produce an extremely fine dispersion of the water phase droplets. Diameters of the droplets at this point range from 2 to 20 μm; the average is about 5 μm. Margarine is a stable emulsion because the continuous phase (the fat) is solid. The chilled emulsion is then held in a quiescent tube, and the margarine is allowed to solidify for a period of time before being extruded to the packaging line. The emulsified water droplets plus a small amount of entrained air provide the desired opacity to margarine.

Whipped Cream and Nondairy Whipped Toppings

Flavorful, airy toppings produced by whipping an emulsion have long been used for desserts by pastry chefs. Foam formation is basically an interfacial phenomenon, but the generation of a good whipped product (including ice cream, which is basically a frozen whipped cream) is somewhat unusual in that whipping involves a controlled deemulsification correlated with crystallization of the fat to stabilize the air cells. The various processes discussed next will be considered from this viewpoint.

WHIPPED CREAM

By centrifugation, the milk fat portion of milk is concentrated to make cream. In the United States, three standards of identity apply to cream:

1. light cream (coffee cream or table cream) contains 18–30% milk fat;
2. light whipping cream (whipping cream) contains 30–36% milk fat; and
3. heavy cream (heavy whipping cream) contains 36% or more milk fat.

The producer may also add approved emulsifiers, stabilizers, nutritive sweeteners, and flavors, if desired. Other countries set their own standards for "cream," so these specifications may be different outside the United States. After concentration, the cream is homogenized, pasteurized, and packaged for the consumer.

To whip properly, the cream must contain sufficient fat (30% minimum), it must be held for at least 24 hr at refrigeration temperatures, and the quantity must not be too large (i.e., the mixer must not be overfilled). The importance of each of these factors will become apparent.

Whipped cream is a foam in which the air bubbles are stabilized by *agglomerated* fat globules. During the early stages of mixing, air is incorporated into the cream and divided into large bubbles. Fat globules concentrate at the air-water interface and stabilize the bubble. (This results from the partial removal of the membrane, exposing some fat and forcing it into the air phase.) At the same time, fat globules in the aqueous phase agglomerate, partly through membrane-membrane interactions and partly through binding of membrane protein to milk proteins. The fat globules must agglomerate but not coalesce (as they do during churning). Thorough cooling of the

Agglomerate—To remain in close proximity but, because of any of a variety of forces, not coalesce; e.g., individual particles in a suspension.

cream solidifies the fat to prevent coalescence. (If the cream is too warm, the result may well be butter rather than whipped cream.)

Most of the air incorporation occurs during the first stage of mixing (before significant viscosity has developed). Further mixing shears and subdivides the air bubbles, and the new air-water interface area is stabilized by the fat globule agglomerates. (Keep in mind that partial removal of the fat membrane exposes hydrophobic fat surfaces that collect at the air-water interface.) Viscosity results from the increase in the number of (subdivided) bubbles and continued agglomeration of fat globules to form a network throughout the system. As viscosity slowly develops, the shear stress on the air bubbles increases, contributing to further subdivision, and the whipped cream rapidly becomes stiff (the end point of the whipping process). If the mixer is overfilled, it takes longer to effect fat agglomeration and thus longer to reach the internal shear stress that contributes to completion of whipping.

Emulsifiers may be added to cream to increase *overrun* and final stiffness. In one study in which combinations of polysorbates were used, it was found that a 50:50 mixture of Polysorbate 60 (the monostearate) and Polysorbate 65 (the tristearate) at 0.5% concentration produced the stiffest cream with an overrun of about 200% (i.e., two volumes of air per volume of cream) (1). Polysorbate 80 (the monooleate) in place of Polysorbate 60 produced much less overrun and a softer whipped cream. Studies in which monoglycerides and two derivatives (the citric and lactic acid esters) were used at the 0.2% level have also been reported (1). Monoglycerides alone had poor results, and a blend of citric acid ester and monoglyceride had good results. Lactylated monoglyceride produced a very stiff foam with a high overrun, which may have resulted from its interfacial film-forming tendencies. The overrun with simple whipping cream tends to be about 100%, and emulsifiers apparently increase air incorporation during the early mixing stages.

NONDAIRY WHIPPED TOPPINGS

Consumers have three concerns about using cream for whipping: the cost, the high level of saturated fatty acids in milk fat, and the cholesterol content. These concerns were addressed by food technologists, who developed whipped toppings made with vegetable fats. Unlike cream, these fats do not have a natural emulsifier, so the necessary emulsifiers are added during production. However, producing a nondairy whipped topping is more complicated than simply combining an emulsified fat with an aqueous protein solution and whipping the mixture. The fat must have the proper solid fat content profile so that the "correct" amount of solid fat is present to stabilize the air bubbles. Also, the emulsifier used must interact with the protein, coating the fat so as to produce the necessary degree of controlled desorption of the protein layer. Finally, the emulsion is frequently spray dried for storage and then reconstituted with cold water before whip-

Overrun—Increase in volume of a whipped material resulting from the incorporation of air.

ping. The emulsifier, protein, and any other ingredients (e.g., sugar, flavors, and maltodextrins) must rehydrate in the proper fashion to produce an emulsion that is whippable. In short, the development of a good nondairy whipped topping is not a simple assignment.

A basic whipped topping powder formula is 25% partially hydrogenated fat and 5% emulsifier melted together and 50% water in which 15% maltodextrin and 5% sodium caseinate are dissolved. The two phases are mixed, homogenized, and then spray dried. The emulsion is stabilized by the combined effects of the emulsifier and protein. The powder is held at 5°C for 1 hr to solidify most of the fat and then stored at temperatures below 20°C.

For whipping, the powder is dispersed in an equal weight of cold water. Agitation initiates many of the same events that occur during whipping of dairy cream—fat globule agglomeration, partial desorption of protein from the fat-water interface, air incorporation (stabilized by fat globules), and air subdivision resulting in formation of a stable foam.

The nature of the fat seems to be crucial. Best results are obtained with partially hydrogenated lauric fats (e.g., coconut or palm kernel fat) plus partially hydrogenated vegetable oils (e.g., soybean or sunflower oil). The hydrophobic segments of the protein, which penetrate the fat globule, appear to inhibit crystallization of the fat, so the fat in the powder itself is a *supercooled* phase. Upon desorption (which is aided by the presence of emulsifier), the fat rapidly crystallizes. Crystallization concurrent with protein desorption appears to be necessary for proper agglomeration and concentration of the fat at the air-water interface, resulting in stabilization of the air bubbles. This phenomenon appears to be connected with short-chain C12 fatty acids; it occurs with lauric fats but not if the fat phase contains solely the C18 vegetable oils. A partially hydrogenated C18 fat, exhibiting no supercooling phenomenon, produces a poor whipped topping, as does a partially hydrogenated C12 fat with a low melting point (little or no fat crystal formation occurs at the temperature of the mixture being whipped).

α-Tending emulsifiers such as propylene glycol monostearate or lactylated monostearin have been used with good success, and Polysorbate 60 (the monostearate derivative) is frequently used in commercial nondairy whipped toppings. Both types form a rather thick layer at the fat-water interface. The α-tending emulsifiers form a multilayer of emulsifier, and Polysorbate 60 forms a layer of adsorbed water (held by the polyoxyethylene chain). These properties promote partial protein desorption, and yet a "sticky" surface is maintained on each fat globule that enhances agglomeration.

Ice Cream

The basic interfacial phenomena that occur during the manufacture of ice cream are similar to those that occur during whipping.

Supercooled—Pertaining to a solution that is cooled below the temperature at which crystals would ordinarily begin to form but do not because of the absence of nucleation or the presence of crystal inhibitors.

Again, the aim is the subdivision of air bubbles and stabilization by adsorbed fat globules. However, in ice cream production, the temperature regime is quite different, so one would expect differences in the emulsifiers used.

After the initial ice cream emulsion is prepared, it is aged for 4–24 hr at ~5°C. This is an important step. During the aging, partial desorption of the fat globule membrane takes place and the globules agglomerate. Desorption is aided by the presence of emulsifiers. As in whipped toppings, it appears that the fat is supercooled, and solidification is connected with the desorption. During agitation and freezing, the fat globules collect at the air-water interface and stabilize the air bubbles. Studies have shown that increasing the degree of desorption (by the addition of a monoglyceride) gives a "drier" ice cream with smaller average air bubble diameter.

The emulsifiers commonly used in ice cream manufacture are monoglyceride (mono- and diglycerides seem to work equally well) and Polysorbate 80 (the monooleate derivative). The effectiveness of the unsaturated emulsifier compared with that of the saturated variety used in whipped toppings is undoubtedly connected to the much lower processing temperature. The emulsifiers are used at a 0.2–0.4% level (compared with the level of 3–5% in whipped toppings). Again, these details relate to the temperature differences. Gums (e.g., guar, carrageenan, and tragacanth) are also frequently used in ice cream mixes. They improve the body and mouthfeel of the product, but their main purpose is to act as crystal modifiers. They inhibit the growth of ice crystals and lactose crystals in frozen ice cream, preventing the gritty mouthfeel that can develop, particularly when the product is stored for a long time.

Coffee Whiteners

Cream in coffee is enjoyed by many people. Today, however, the cream pitcher has disappeared from the table, replaced by a powder, which when stirred into a cup of coffee, mimics the effect of cream. The restaurant trade uses small, individual packets of liquid whitener, while the powder (a spray-dried emulsion stable at room temperature) is more popular for home use.

A wide range of formulations appears in the literature, but a typical spray-dried whitener contains vegetable fat, 37%; corn syrup solids (dextrose equivalent, 42), 56%; sodium caseinate, 5%; dipotassium phosphate, 1.6%; monoglyceride (e.g., glycerol monostearate), 0.3%, Polysorbate 65, 0.1%; and flavor and anticaking agent as desired. The emulsifiers are added to the fat, the other materials are dissolved in water, and an emulsion of the two phases is homogenized and then spray dried. The powder must be held under cool conditions for a tempering period so that the fat solidifies and clumping is avoided.

The fat for coffee whitener usually has a melting point of about 42–45°C, but it has a rather steep solid fat content profile. Such a fat melts completely at coffee temperature (thus leaving no waxy mouthfeel) but is relatively solid at room temperature, so the powder does not clump during storage. Corn syrup solids provide body during processing, and the dried matrix separates the fat particles.

The added emulsifier serves two purposes: during homogenization, it facilitates formation of fine fat globules (average diameter, 1 µm), and in conjunction with the protein, it prevents *feathering*, which occurs if the emulsion breaks down and the fat separates (oils out) when the whitener is dissolved in the coffee. The protein helps emulsify the fat, and the proper emulsifier can stabilize the protein layer when the whitener is added to the coffee. The protein layer can be destabilized by calcium ions and organic acids in the coffee, and dipotassium phosphate counteracts this tendency. One patent recommends the use of sodium stearoyl lactylate rather than Polysorbate 65 because its protein-complexing behavior apparently makes the protein more effective (2).

> **Feathering**—Streaks of fat and precipitated protein that form in a liquid such as coffee when improperly stabilized whitener is added; caused by breakdown of the emulsion.

Troubleshooting

The main role of emulsifiers in ice cream and whipped toppings is to destabilize the natural proteinaceous film surrounding milk fat globules, allowing the solid milk fat to agglomerate around the air bubbles and thus stabilize the three-dimensional physical structure. Interaction of emulsifier with the fat can change the finished properties. One has to keep in mind that defects may not be caused by the emulsifier system and may have other causes.

ICE CREAM		
Symptom	Causes	Changes to Make[a]
Inadequate overrun	Air bubble instability caused by poor fat agglomeration	Use aerating emulsifiers (e.g., GMS) or α-tending emulsifiers (e.g., PGMS).
Too soft; shiny, wet looking	Insufficient agglomeration of fat globules in the freezer	Use emulsifiers containing unsaturated fatty acids (e.g., GMO or Polysorbate 80).
Too firm; "dry" looking	Excessive agglomeration	Adjust emulsifier blend to contain more saturates (e.g., GMS or Polysorbate 65).
Mix viscosity too high	Homogenization temperature too low	Increase homogenization temperature.
	Worn homogenizer valves	Regrind valves.
	Mix acidity too high	Adjust pH.
	Homogenizer pressure too high	Adjust pressure.
Mix viscosity too low	Hold time before freezing too short	Adjust hold period.
Curdled meltdown	Protein destabilization caused by high acidity	Add dipotassium phosphate or sodium citrate.
	Homogenizer pressure too high	Lower homogenizer pressure.

[a] DATEM = diacetyl tartrate ester of monoglyceride; GMO = glycerol monooleate; GMS = glycerol monostearate; PGMS = propylene glycol monostearate; and SSL = sodium stearoyl lactylate.

WHIPPED TOPPINGS		
Symptom	**Causes**	**Changes to Make**
Insufficient overrun	Poor emulsification	Increase the amount of emulsifier with a high hydrophilic/lipophilic balance (e.g., sorbitan monostearate, Polysorbate 60 and/or 65). Check homogenizer valves for wear.
	Whipping temperature too high	Adjust whipping temperature.
Soft whipped product	Poor fat agglomeration	Use an unsaturated fatty acid emulsifier (e.g., GMO) for part of the system. Use an α-tending emulsifier (e.g., PGMS).
	Poor homogenization	Decrease valve clearance. Increase pressure. Make multiple passes.

COFFEE WHITENER		
Symptom	**Causes**	**Changes to Make**
Feathering	Emulsion breakdown	Use SSL or DATEM as the emulsifier.
	Protein destabilization	Add dipotassium phosphate or sodium citrate.
Clumping	Untempered fat	Make sure the heat of fat crystallization is released before packaging.
	Soft fat	Use a fat that remains solid at room temperature.

References

1. Sogo, Y., and Kako, M. 1989. Whipping cream emulsifiers. In: *Food Emulsifiers*. G. Charalambous and G. Doxastakis, Eds. Elsevier, Amsterdam.
2. Miller, D. E., and Werstak, C. E. 1983. U.S. patent 4,415,600.

Supplemental Reading

Chandan, R. 1997. *Dairy-Based Ingredients*. American Association of Cereal Chemists, St. Paul, MN.

CHAPTER 6

Dressings and Sauces

Dressings and sauces are a part of traditional cuisines of all types. In terms of interfacial chemistry, they range from the very simple (e.g., an oil and vinegar salad dressing) to the complex (e.g., mayonnaise). A common ingredient is oil, which gives a smooth mouthfeel and extended flavor impact to the food. To function effectively, the oil should be present in a finely divided (i.e., emulsified) state. The means for accomplishing this are many and often involve materials other than (or in addition to) the emulsifiers discussed in Chapter 3. These materials may be proteins (see Chapter 3) or various *polysaccharides*.

The polysaccharides are often referred to as "stabilizers," and they work in different ways to maintain the oil in small droplets. Some *gums*, for example, actually adsorb to the oil-water interface, forming a film with good interfacial viscosity and preventing droplet coalescence. Other polysaccharides inhibit creaming (see Chapter 1) of the emulsified drops by greatly increasing the viscosity of the aqueous phase, while still others prevent creaming by gelling, i.e., forming a semisolid matrix that prevents flotation of the oil drops. Thus, in the strict sense of the word, few of these materials are true "emulsifiers" (surfactants that lower interfacial tension), but they still perform the needed function of stabilizing the dressing or sauce by preventing separation of the oil phase.

> **In This Chapter:**
>
> **Polysaccharides at Interfaces**
>   Gums
>   Modified Starch
>   Cellulose Derivatives
>
> **Salad Dressings**
>   Pourable Salad Dressings
>   Spoonable Salad Dressings
>
> **Mayonnaise**
>
> **Reduced-Fat Dressings and Sauces**
>
> **Troubleshooting**

Polysaccharides at Interfaces

GUMS

Obtained from *plant exudates*, seaweed extracts, and bacterial polymeric products, gums are high molecular weight polysaccharides that dissolve in water to form viscous solutions and, in some cases and under the proper conditions, gels. They are widely used in food products, usually at low levels (0.1–1%), for many different functional reasons. Some gums have a high water-binding capacity and are used to control water migration in the finished product. These gums also act to inhibit growth of ice crystals in frozen products. In other instances, the viscosity imparted by the gum is necessary for the intermediate material (e.g., a cake batter) to perform properly when baked or cooked. Gum viscosity is often used as a means of classification.

Polysaccharide—A carbohydrate containing several hundred, thousand, or hundred thousand sugar units (from the Greek *poly,* meaning "many").

Gum—Polysaccharide that markedly increases viscosity when dissolved or dispersed in water.

Plant exudate—Droplets of hardened gum exuded by a plant to seal a break in the bark.

TABLE 6-1. Viscosities of Gums

Gum	Viscosity[a]
Low viscosity	
Arabic	2–5
Ghatti	4–10
Larch	2–10
Medium viscosity	
Sodium alginate	25–800
Propylene glycol alginate	100–500
Tragacanth	200–500
Xanthan	800–1,400
High viscosity	
Guar	2,000–3,500
Karaya	2,500–3,500
Locust bean	3,000–3,500
Cellulose gums	
Sodium carboxymethylcellulose	50–5,000
Hydroxypropyl methylcellulose	20–50,000
Methylcellulose	10–2,000
Gel-forming gums	
Agar	Gel
Calcium alginate	Gel
Carrageenan	Gel
Furcelleran	Gel
Gellan	Gel
Pectin	Gel

[a] Viscosity of a 1% aqueous solution of the gum, in centipoise. (Water viscosity = 1 centipoise.)

Oligosaccharide—Short polymer of sugar having three to eight sugar units.

Uronic acids—Derivatives of a sugar, in which the terminal -CH$_2$OH group is oxidized to a carboxylic acid. D-Glucuronic acid is derived from D-glucose, D-mannuronic acid from D-mannose, and D-galacturonic acid from D-galactose.

Hexose—Sugar containing six carbon atoms.

Pentose—Sugar containing five carbon atoms.

Common food-grade gums and the viscosities of 1% solutions are listed in Table 6-1.

The molecular structure of gums is a long polysaccharide chain with numerous side branches of sugars or *oligosaccharides*. Frequently, the sugar units include carboxylic (*uronic*) acids, e.g., D-glucuronic, D-mannuronic, or D-galacturonic acid. In a few instances, sulfate esters provide an anionic character. Many different saccharides are found in gums, including the *hexoses* D-glucose, D-mannose, and D-galactose and the *pentoses* D-arabinose, D-xylose, and D-rhamnose. The highly branched structure contributes to water solubility, and the anionic gums often form gels in the presence of cations such as Ca^{++}.

The low-viscosity gums (arabic, ghatti, and larch) are used mainly as water-binding agents; e.g., they prevent *syneresis*, or weeping, in baked meringues. The gums readily dissolve in water, and a concentrated solution (10–50%) of the gum is often used to emulsify hydrophobic flavor oils such as citrus or cinnamon oil. The gum forms a layer at the oil-water interface, and spray drying the emulsion results in an encapsulated flavor that has many uses in food product development.

The medium-viscosity gums (sodium alginate, propylene glycol alginate, tragacanth, and xanthan), when used in foods at the usual levels, impart *body* to the product. They also have emulsifying properties and are often found in pourable dressings, e.g., oil and water types that are emulsified just before use by shaking the bottle. They can also be used in flavor oil emulsions, particularly the liquid types often used in bakeries.

High-viscosity gums (guar, karaya, and locust bean) are used as thickening or stabilizing ingredients. They greatly increase the viscosity of the aqueous phase of the food product and aid in air incorporation in whipped toppings. High-viscosity gums are often used in low-fat or fat-free pourable dressings. The enhanced viscosity (or body) results in a mouthfeel that is similar to that of fat-containing dressings.

Gel-forming gums (agar, calcium alginate, carrageenan, furcelleran, gellan, and pectin) are used mainly in foods such as jellies and fruit fillings that require a semisolid nature. These gums also impart *freeze-thaw stability* to many products. Many frozen whipped toppings and ice creams include one or more of the gums listed, particularly carrageenan or alginate.

Most dressings and many sauces contain vinegar and thus are acidic. The effect of pH on gums is quite variable, and suppliers can provide information on the performance of gums under a range of pH conditions. Some gums are somewhat susceptible to acid hydrolysis, but again this varies (tragacanth is reasonably stable at pH 2, whereas arabic begins to depolymerize at pH 4). Most gums (e.g., the carboxylic acid groups or sulfate esters) are acidic in nature, and a certain sensitivity to pH would be expected. For example, the viscosity of a sodium alginate solution is constant from pH 4 to 10, but at lower pH, the carboxylate groups are converted (un-ionized) to carboxylic acids and the polymer chains begin to interact, resulting in an increase in viscosity. Propylene glycol alginate ester, on the other hand, is relatively insensitive to pH, and the viscosity increase at lower pH is much less than that of sodium salt.

MODIFIED STARCH

As described in Chapter 4, starch is made of two types of molecules: linear amylose and branched amylopectin. In native starch, these molecules exist partly as amorphous chains and partly in crystallites (regions of crystallized starch). When starch is heated with a liquid, gelatinization occurs. The starch first hydrates (swells), and then the crystallites melt into the water solvent. Upon cooling, retrogradation (recrystallization) of the gelatinized starch takes place, and a gel is formed, the properties of which depend upon the starch-water ratio of the mixture and the amylose-amylopectin ratio of the starch. Starches can be modified to make them more useful in the food industry.

Crosslinked starch is starch that has been treated with one or more reagents to form bonds between glucose residues in adjacent starch chains. Sodium trimetaphosphate or phosphorus trichloride form a phosphate diester:

2 Starch-OH + PO_3Cl (NaOH) → Starch-O-P(=O)(-O^-)-O-Starch Na^+

Epichlorohydrin forms a diether bridge, where the central moiety is a hydroxypropyl group:

Starch-O-CH_2-CH(-OH)CH_2-O-Starch

The differences in reactivity of phosphate esters and ethers lead to differences in the chemical stability of the two kinds of modified starch. They both are generally useful, and each has advantages in certain situations, such as extremes of pH and temperature.

Crosslinking raises the gelatinization temperature of starch, but more important, the gel that is formed is stable under varying conditions of temperature, low pH, and shear. Of course, the degree of stabilization depends on the degree of crosslinking (i.e., the amount of reagent used during the modification reaction). Crosslinked starch is less likely to lose viscosity at low pH than other kinds of starch. This makes crosslinked starch useful in salad dressing, in which the pH is

Syneresis—Separation of a liquid from a gel; weeping.

Body—A qualitative measurement determined in sensory tests as the ability to "fill the mouth" with the characteristic being measured.

Freeze-thaw stability—The ability of a product to be frozen, thawed, and refrozen several times without noticeable changes in physical characteristics.

Crosslinked starch—Starch in which hydroxyl groups on adjacent chains are joined by a covalent linkage.

Stabilized starch—Starch in which hydroxyl groups form ester or ether bonds with other small molecules.

Shear thinning—Decrease in the viscosity of a suspension as shear rate increases.

Crystal inhibitors—Small molecules that interfere with the deposition of dissolved molecules on the growing face of a crystal, thus inhibiting growth of the crystal.

3–4. However, the starch degrades somewhat when it is cooked at low pH, so it is precooked (gelatinized) at neutral pH before it is combined with the oil, vinegar, and spices.

Stabilized starch is derivatized without crosslinking. Reaction with propylene oxide, for example, produces hydroxypropylated starch. Sodium tripolyphosphate, under the proper reaction conditions, makes phosphate monoester derivatives, and acetic anhydride yields acetylated starch. Stabilized starches have lower gelatinization temperatures and higher viscosities than the native starches, but they are often less resistant to *shear thinning*. A major advantage of stabilized starches is that because of the subtitution along the starch chains, they are less inclined to retrograde upon cooling. Retrogradation renders the starch opaque (rather than translucent) and much firmer, possibly generating a gritty or rubbery mouthfeel.

CELLULOSE DERIVATIVES

Cellulose is insoluble in water, but substitution of certain groups along the glucose backbone render it soluble. The three forms most commonly used in foods are substituted with carboxymethyl, methyl, and hydroxypropyl plus methyl groups. The degree of substitution varies depending upon the cellulose-reactant ratio, and products with a wide range of viscosities are possible. Methylcellulose has a modest degree of surface activity and acts as an emulsifier in oil-in-water systems. Hydroxypropyl methylcellulose forms a film at oil-water interfaces and stabilizes oil emulsions. Certain grades of hydroxypropyl methylcellulose form gels even at room temperature and can be used in some types of fat-free pourable dressings to add body.

Microcrystalline cellulose is a highly purified fraction of regular cellulose. It absorbs several times its own weight of water and is added to some low-fat dressings for body. A suspension of microcrystalline cellulose is thixotropic; i.e., it flows very slowly at low shear rates, or has a high apparent viscosity. For this reason, it is often added to condiment sauces (e.g., ketchup and barbecue sauce) to enhance "cling"; i.e., the sauce remains where it is applied and does not run down the sides of the food item.

Salad Dressings

POURABLE SALAD DRESSINGS

According to U.S. standards of identity, french dressing is a mixture of vegetable oil (minimum 35% by weight), acidifying ingredients (e.g., vinegar, lemon juice, or lime juice), and other permissible ingredients (salt, sugars, spices, monosodium glutamate, tomato products, eggs, colorants, thickeners, citric and/or malic acids, sequestrants, and *crystal inhibitors*). Thus, a wide range of products is included, from the simplest vinaigrette (oil, vinegar, and a few spices) to relatively complex products marketed under a variety of names.

Salad dressing oil is a refined, bleached, deodorized oil that has been *winterized* to inhibit development of cloudiness during refrigeration. A crystal inhibitor is often added to lengthen the time that elapses before such cloudiness begins to develop. The molecular structure of crystal inhibitors is similar to that of triglycerides but different in some specific manner. When the crystal inhibitors deposit on the faces of growing fat microcrystals, this difference in molecular structure interferes with the further deposition of fat molecules. Two compounds, oxystearin and polyglycerol esters, are approved by the U.S. Food and Drug Administration specifically for this use. Oxystearin is made by blowing air through cottonseed or soybean oil (hydrogenated to an iodine value of about 35) at 200°C. The product contains many polymeric and breakdown products, and its exact composition is unknown. Its maximum permissible use concentration is 0.125% by weight. Several emulsifiers, including sucrose esters, glucose esters, and sorbitan tristearate, also inhibit crystal formation.

Before a dressing is added to the salad greens, the bottle is shaken to make an emulsion, i.e., an even distribution of the oil and water phases. Many nonstandard dressings are rather complex mixtures that also contain dairy products (e.g., buttermilk powder or blue cheese), spices (e.g., ground mustard or turmeric powder), dried vegetable pieces (e.g., onion or green and red peppers), and other items. They also often include Polysorbate 60 in their formulations at the maximum permissible level of 0.3% by weight of the total dressing to enhance "home emulsification." An emulsion stabilizer helps maintain uniformity of this complex mixture and increases the product appeal.

The viscosity of a simple emulsion of 35% oil in water is almost that of water. For many dressings, a higher viscosity is desired so that the dressing does not drain quickly from the salad greens and collect in the bottom of the salad bowl. Soluble gums are used at a concentration of 0.05–0.3% to give the desired viscosity. Some of the names commonly seen on labels include propylene glycol alginate, xanthan gum, modified cellulose gum, and carrageenan. Besides contributing the desired viscosity, the gum must be stable in an acidic environment; hydrolysis at low pH decreases viscosity.

Microcrystalline cellulose may be used at 1–2% concentration and increases viscosity by decreasing the amount of continuous (water) phase in the formulation. It is most often used in dressings that contain tomato products, where it gives a smooth texture and imparts a certain degree of thixotropy to the mixture so that the dressing clings to the surface of the salad greens.

SPOONABLE SALAD DRESSINGS

Spoonable, starch-based salad dressing was first developed as a low-cost alternative to mayonnaise. Today, it is accepted as a somewhat different product, purchased and judged by consumers on its own merits. While it is similar to mayonnaise in rheology (it is a *Bingham plastic,* and its yield value is a factor in acceptability), its characteris-

Winterize—To cool salad oil until high-melting-point triglycerides form crystals. These crystals are removed so that the next time the oil is cooled, the cloudiness that results from crystallization will not occur.

Bingham plastic—A suspension that does not flow when subjected to a shear stress lower than the yield value but does flow at any shear stress above the yield value.

> **Box 6-1. Yield Value**
>
> Spoonable dressing is a "Bingham plastic" in rheological terms (Fig. 6-1). When shear stress (stirring) is applied at low levels, the dressing acts like a solid; i.e., it does not flow, although it may fracture (break). At some level of shear force, called the "yield value," the dressing begins to act like a liquid. Put another way, if a block of dressing is simply cut, the vertical edge does not flow, because the shear stress of gravity is less than the yield value. However, the dressing can be spread smoothly with a knife on a slice of bread because the shear force applied by the knife blade is greater than the yield value. The yield values of spoonable dressing and mayonnaise are important factors in the quality of the products.

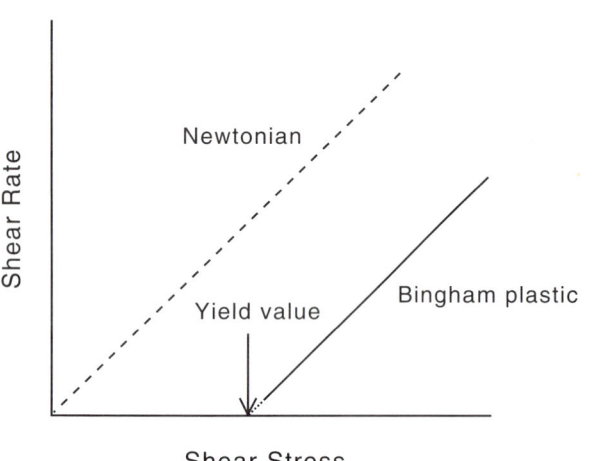

Fig. 6-1. Rheological stress-strain curve of a Bingham plastic.

Waxy maize starch—Starch from genetically modified corn; contains almost no amylose.

Lipoprotein—A complex of a protein with a lipid (generally a phospholipid) found in many plants. The most common in food processing is from egg yolk.

tics are dependent on the nature of the food starch used and not on the state of emulsification of the oil.

A salad dressing meeting the U.S. standard of identity contains vegetable oil (minimum 30% by weight), egg yolk (minimum 4% by weight), acidifiers (vinegar, lemon juice, or lime juice), a paste prepared from a suitable food starch, and other ingredients (salt, sugar, spice, monosodium glutamate, thickeners, citric and/or malic acid, sequestrants, and crystal inhibitors). A typical commercial product might contain (by weight) 35% salad oil, 4% liquid egg yolk, 2% salt, 11% sugar (or 15.5% high-fructose corn syrup), 0.5% mustard flour, 11% vinegar (100 grain), 6% modified starch, spices, and water to total 100%.

The starch is the key component in this formulation, providing the desired structure and creamy texture to the finished product. The starch most often used is a *waxy maize starch,* which is essentially 100% amylopectin, crosslinked with sodium trimetaphosphate and stabilized, usually by hydroxypropyl substitution. The starch has a relatively high gelatinization temperature and sets to a soft gel upon cooling. The crosslinking prevents hydrolysis in the low pH environment of the finished dressing; without crosslinking, the starch gel would soften during storage. The stabilization interferes with recrystallization of the side chains and maintains a creamy texture in the finished dressing during storage, particularly if the dressing is held at refrigerator temperatures or even frozen.

Spoonable salad dressing is made in several stages. First, the oil is emulsified with the egg yolk and some of the water and vinegar. The egg yolk *lipoproteins* are the surface-active materials that stabilize this emulsion, which is rather coarse because of the relatively large diameter of the oil drops. Meanwhile, the starch is gelatinized by heating it with water. After it is cooled, the other ingredients such as vinegar, sugar, salt, and spices are mixed in to form a paste. The cooled paste is then fed into the premixer and blended into the oil-egg emulsion. This soft mass is pumped to a *colloid mill,* where the final high viscosity and smooth texture are generated. The gelatinization procedure is crucial in attaining good final product quality, and the details of time and temperature depend to a large degree upon the specific starch used. Information about the proper operation of the starch cooker is best obtained from the starch supplier.

Mayonnaise

Standardized mayonnaise contains vegetable oil (minimum 65% by weight), acidifiers (vinegar, lemon juice, or lime juice), egg yolks (liquid, frozen, or as whole egg), and other ingredients (salt, sugars, spices, monosodium glutamate, sequestrants, citric and/or malic acid, and crystal inhibitors). Mayonnaise is an oil-in-water emulsion, stabilized by the lipoprotein components of egg yolk. While the legal minimum for the oil content is 65%, mayonnaise at this oil level is rather thin (i.e., has low viscosity). Thus, the usual commercial product today contains 77–82% oil. Liquid egg yolk (45% solids) is the emulsifier and is used at 5.3–5.8% of total formula weight. Sometimes whole eggs (25% solids) are substituted for egg yolks on a total solids basis (i.e., whole eggs at 9.5–10.4% of the total formula weight). This produces a somewhat "stiffer" product than egg yolks because the egg albumin is denatured at the interface and forms a matrix in the aqueous phase, increasing the Bingham yield value of the product.

Oil is the internal phase in the emulsion. If all the droplets are spherical and incompressible and have the same diameter, the maximum volume percentage of oil is 74.05%. (For example, if a 1,000-ml container were filled with small, uniform ball bearings, 260 ml of water would be needed to fill the space between the bearings.) If the oil volume percentage exceeds 74%, the emulsion inverts (i.e., the oil droplets coalesce, and the emulsion becomes a water-in-oil emulsion). However, if the droplets differ in diameter, small droplets can fill the spaces between large drops. This is what happens in mayonnaise.

This "filling in the spaces" also has a major effect on the rheological (flow) characteristics of the product. Mayonnaise is a Bingham plastic, and the yield value is related to the amount of internal (oil) phase in excess of the 74% (by volume) theoretical limit. Thus, a mayonnaise with the legal minimum of 65% oil by weight has a very low yield value and is considered too "thin" by users. Mayonnaise with 80–84% oil by weight has a high yield value and is considered dry or rubbery by home consumers, although it is preferred for institutional use because it does not soak into bread in sandwiches or soften and flow over salads.

Colloid mill—Machine used to decrease the size of suspended particles or droplets.

As mentioned above, the lipoproteins in egg yolk are the main emulsifying agents in mayonnaise. They are partially denatured by the low pH of the vinegar used, and their emulsification capabilities are enhanced. If whole eggs are used, the egg white albumins are also denatured by the low pH and provide some emulsification, giving the aqueous phase a somewhat meringuelike quality. The resulting mayonnaise may be slightly fluffier than mayonnaise made with only yolk, and the yield value is slightly greater. The choice between the two products is mainly a matter of consumer preference.

An oil that resists clouding is essential for mayonnaise. After the jar is opened, it is stored in the refrigerator. If fat crystals form, the balance of emulsion-stabilizing forces is disrupted and the emulsion breaks down. In addition to winterization of the oil, crystal inhibitors may be used to further inhibit clouding.

The preparation of mayonnaise is difficult, and science is less important than the experience and skill of the equipment operators in making an acceptable final product. Nevertheless, there are certain basic principles that apply. Understanding them can help to solve production problems when they arise.

A planetary mixer (e.g., a Hobart) with a paddle is used in traditional mayonnaise production. The yolk and other dry ingredients are blended, and then the oil is added while mixing continues, slowly at first and then more rapidly when the mass begins to thicken (without allowing any puddles of oil to form). After all the oil is incorporated, the vinegar is added and the mixture is thoroughly blended. This method is used in homes and in gourmet restaurants that have their own special blends.

Commercial manufacture is a two-step process. First, a premix similar to that in the traditional method is made. Often mayonnaise from a previous batch is placed in the mixer to give a high-viscosity medium into which the mixer blades can readily disperse the oil. The egg and dry ingredients are added, mixing is initiated, and the oil and vinegar are pumped in at a controlled rate (the flow of vinegar is not completed until after all the oil is in). The viscosity of the premix depends in part upon the characteristics of the particular batch of egg being used and can be adjusted by changing the timing and rate of the oil and vinegar addition. In the second step, this rather soft premix is pumped to a colloid mill, which reduces the diameter of the oil droplets. The viscosity of the final product depends upon the degree of subdivision attained in the mill, which in turn depends upon mill settings (e.g., clearance between rotor and stator and speed of the rotor) and the premix viscosity. The ideal, of course, is to obtain a high degree of subdivision but large variation in droplet diameters. If too much subdivision occurs and all droplet diameters approach the theoretical minimum (governed by interfacial tension), variable space filling is lost, and since the volume percentage of oil is greater than 74%, the emulsion inverts (breaks down). As stated earlier, the ability of the operator to adjust the mill to achieve the proper degree of subdivision depends more on skill and experience than on science.

Reduced-Fat Dressings and Sauces

The first reduced-fat dressing was a spoonable, starch-based product developed during the 1930s as a less expensive alternative to mayonnaise. With less than half the oil of mayonnaise, this product would qualify for a "reduced fat" label today, although that was not the original intent of the developers. The appearance of "low-fat" and "fat-free" salad dressings (both pourable and spoonable types) during the 1990s has been in response to the consumer trend toward lowering fat in the diet. This trend is not confined to dressings, of course. Bakery, dairy, confectionery, and numerous other food products have also been developed.

The challenge in all areas is to mimic the functionality of fat while reducing its level in the product. Since fat contributes many of the positive gustatory characteristics of foods, this has not been easy to do. In dressings, for example, some of the major contributions of fat are body (viscosity and cling), creaminess (smoothness), lubricity (slipperiness), sheen (appearance), and flavor (intensity and duration). The search for nonfat materials to mimic each of these features has produced some successes but no one (or even two) materials that do the whole job.

Emulsified oil increases the viscosity of the dressing, although the increase is not great, except in mayonnaise. Still, this viscosity increase is desirable because it slows drainage of the dressing from the salad greens. Gums, modified starches, and solid fibers all increase viscosity. By themselves, however, they do not make a satisfactory product. Some gums, for example, increase viscosity, but the dressing has a "slimy" rather than a smooth or slippery mouthfeel. The most popular gums used today in low-fat or fat-free dressings are xanthan, propylene glycol alginate, and hydroxypropyl methylcellulose (labeled simply as cellulose gum). They are used in part because of their high resistance to hydrolysis during storage at the low pH of the dressing. While these materials do improve cling, that alone is not sufficient.

A *fat mimetic* comes closer to providing the needed creaminess and lubricity. The first successful one was Simplesse, a product made from egg and dairy proteins in the form of spherical particles about 0.5–2 mm in diameter. The claim was that these gave a "ball bearing" effect on the palate. Soon after, starch gels containing soft (deformable) particles appeared that produced a similar creamy effect.

The sheen caused by a thin layer of oil is desirable in certain products, particularly condiment sauces. This appearance has not been achieved with starches, gums, or protein derivatives. However, by using a strongly surface-active emulsifier, such as one of the polysorbates, a very small amount of oil is enough to produce this surface effect. If the level of oil plus emulsifier is such that a serving of the product contains less than 0.5 g of fat (by analysis), it may be labeled "fat free."

Fat mimetic—Fat substitute that mimics the properties of fat.

Flavor notes—Small molecules, usually volatile hydrocarbons, that interact with taste receptors and are experienced as flavors.

The main problem with reducing or removing fat is in the area of flavor. Fat plays an important role in the perception of flavor. It modifies the way *flavor notes* interact with taste receptors on the tongue. It influences the order of release of these flavors and slows down the rate of interaction. Since many of the flavor notes derived from spices are fat soluble, they are affected by these factors. A fat-free, vinegar-based dressing containing spices (e.g., oregano, marjoram, and sage) has a large, immediate flavor impact that rapidly fades, leaving only the acidity of the vinegar on the palate. The corresponding dressing with oil has a longer, less intense flavor impact. The flavor notes interact with the sourness in a more pleasing sequence, and the consumer has a more favorable final perception. In the flavor chemist's terms, the result is a more "rounded" flavor. Also, the oil tends to mask low-level off-flavors and mitigates unfavorable interactions between flavor notes. Many flavor molecules are volatile hydrocarbons and are rather unstable in an aqueous environment.

This entire area is still under investigation. Simply using the same spices and other flavors in a fat-free version of an existing regular dressing is inadequate. The formulator must do an extensive rebalancing of the formula to achieve an acceptable result. In dressings, as in the other food applications, experience is showing that low-fat or reduced-fat formulations are often much more acceptable to the consumer than fat-free products and can contribute to bringing the calories from the fat portion of the diet below the recommended 30% of total calories.

Troubleshooting

SPOONABLE DRESSINGS		
Symptom	**Causes**	**Changes to Make**
Oil separates during storage	Crystal formation in oil	Use a winterized (non-clouding) oil. Add a crystal inhibitor to the oil.
	Stabilizer breakdown due to low pH	Use a stabilized (hydroxypropyl) starch. Use propylene glycol alginate gum.
	Insufficient stabilizer	Add a solid particle stabilizer such as mustard flour.
Breaks down during processing	Excess shear force	Increase clearance in homogenizer. Decrease homogenizer rotor speed. Use shear-resistant starch.

POURABLE DRESSINGS		
Symptom	Causes	Changes to Make[a]
Oil separates too quickly	Inadequate emulsification	Add emulsifier (e.g., Polysorbate 60 up to 0.3% or DATEM).
		Use some protein stabilizer (e.g., whey isolates).
	Low viscosity	Add a pH-stable, high-viscosity gum (e.g., hydroxypropyl methylcellulose).
Lacks "cling"	Low viscosity; poor flocculation	Add a fibrous thickener (e.g., microcrystalline cellulose or xanthan gum).

REDUCED-FAT DRESSINGS		
Symptom	Causes	Changes to Make
"Slimy" mouthfeel	Viscosity too high	Switch to lower viscosity gum stabilizer.
Evanescent flavor impact	Insufficient lipid phase	Rebalance flavor system.
		Add emulsifier(s) that form a mesophase or micelles (e.g., SSL + GMO).

[a] DATEM = diacetyl tartrate ester of monoglyceride; GMO = glycerol monooleate; and SSL = sodium stearoyl lactylate.

Supplemental Reading

Whistler, R. L., and BeMiller, J. N. 1996. *Carbohydrate Chemistry for Food Scientists.* American Association of Cereal Chemists, St. Paul, MN.

CHAPTER 7

Beverages

In This Chapter:

Flavor Emulsions
  Oil Phase
  Emulsion Stabilizers
  Emulsion Preparation

Stability
  Creaming
  Flocculation
  Coalescence

Microemulsions

Troubleshooting

Soft drinks are consumed by people around the world. The most popular flavors are citrus (orange first, followed by lemon). In the United States, cola flavor holds the number one spot, followed by orange and lemon. Citrus flavors are *essential oils* extracted from the fruit peel and are insoluble in water. The beverages made with these oils consist of dilute oil/water (O/W) emulsions in dilute solutions of sugar (which may be carbonated). At the bottling plant, a concentrated emulsion is mixed with the aqueous phase, packaged in bottles or cans, and delivered to the vendor. The dilute emulsion must remain stable for at least six months and often must survive intermediate storage under either warm (>40°C) or cold (<10°C) conditions.

A characteristic desired in some beverages is cloudiness. For these drinks, the emulsion includes a flavorless oil called a *weighting agent*, which has a density higher than that of the flavor oil. Adjustments are made so that the density of the oil droplets approaches that of the dilute sugar solution, keeping the emulsion droplets uniformly distributed throughout the beverage during storage and consumption.

Flavor Emulsions

OIL PHASE

Citrus flavor oils are mainly mixtures of *monoterpenes* (typically more than 90%) and *sesquiterpenes*. By themselves, these hydrocarbons are nearly odorless and flavorless, but they are carriers for the compounds—alcohols, aldehydes, ketones, acid, and esters—that provide the citrus aromas and flavors of orange, lemon, grapefruit, etc. The selection and blending of citrus oils to produce the desired flavor and odor is done by flavor chemists and is outside the scope of this book. It is a highly developed expertise and is basic to the design of a product that is acceptable to consumers.

The terpenes have specific gravities of 0.845–0.890 g/cm^3, well below the specific gravity of 1.04–1.048 g/cm^3 of the 10–12% sugar solution of the beverage. The need for inclusion of a weighting agent to make an emulsion with a neutral density (one that does not cream) is apparent. However, the selection of a weighting agent has been complicated by government regulations in various countries. Some agents are permitted in one country but not in another. Thus, no single weighting agent is universally applicable.

Essential oils—Aromatic and flavorous oils obtained from various plant sources and originally known as "essences," mainly in the perfume industry.

Weighting agent—An oil-soluble material that has a density greater than that of water.

Monoterpenes—Aliphatic compounds, containing 10 carbon atoms, formed biologically by combination of two molecules of isoprene (2-methyl butadiene), with subsequent modification to give the various members of this group.

Sesquiterpenes—Similar to monoterpenes but containing 15 carbon atoms; formed by reaction of three isoprene units.

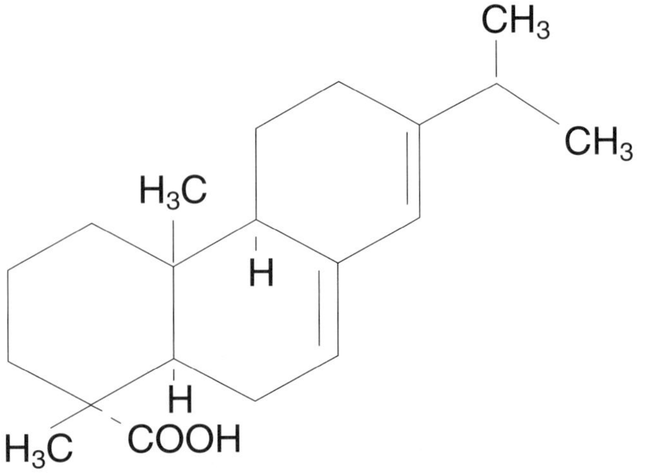

Fig. 7-1. Structure of abietic acid, one of the main diterpenoid carboxylic acids present in purified wood resin.

Brominated vegetable oil—Oil made by treating an unsaturated vegetable oil with molecular bromine. Addition of HBr across double bonds gives these oils a high density.

Ester gum—Gum made by esterifying the acids of purified wood resin with glycerol.

Damar gum—A shrub exudate consisting mainly of triterpenes (containing 30 carbon atoms) but with numerous other related compounds also present.

Sucrose acetate isobutyrate—The octaester of sucrose with acetic and isobutyric acids.

Diterpenoid carboxylic acids—Tricyclic compounds containing 20 carbon atoms and with a carboxylic acid function attached.

Brominated vegetable oil (BVO) was first used in the 1940s. It has a specific gravity of about 1.33 g/cm^3, the highest of the agents presently in use, and thus is the most efficient. Unfortunately, in 1970, several European countries withdrew permission for its use, and shortly thereafter, Canada and the United States limited it to a maximum level of 15 ppm in beverages, about one-tenth that necessary to make a neutral-density emulsion. These actions presented the industry with a major problem. The weighting agents most used today are *ester gum*, *damar gum*, and *sucrose acetate isobutyrate,* although none of them are approved for use in all countries.

Ester gum is produced by esterification of wood resin with glycerol. Wood resin is obtained by purification of the oleoresin from pine trees and consists mainly of *diterpenoid carboxylic acids,* e.g., abietic acid (Fig. 7-1) and pimaric acid. Esterification with glycerol produces a mixture of the mono-, di-, and triglycerides. After purification (by steam sparging and vacuum distillation), a flavorless, colorless, oil-soluble resin is obtained. Since a typical density of this resin is 1.08 g/cm^3, the oil phase requires a much higher proportion of this product than of BVO to obtain neutral density. Ester gum is approved for use in the United States at a maximum level of 100 ppm but is not approved in Canada or most European countries.

Damar gum is a resinous plant exudate from shrubs belonging to the genus *Shorea*. It is a complex mixture of triterpenes, other low molecular weight polymers, and numerous acids, esters, and ketones derived from these compounds. The crude exudate is deodorized by solvent extraction and/or fractional distillation to remove low molecular weight terpenes that might impart a flavor or odor to beverages. Damar gum is completely soluble in essential oils and has a typical specific gravity of 1.05–1.08 g/cm^3. It is permitted for use in European countries as an all-natural, chemically unmodified, vegetable gum resin. It is not approved in the United States for use as a food additive.

Sucrose acetate isobutyrate is a mixture of sucrose esters, mainly the diacetate hexaisobutyrate isomer. It is prepared by esterification of sucrose with the two acid anhydrides, acetic anhydride and butyric anhydride, followed by purification. Its specific gravity is 1.146 g/cm^3, higher than that of ester gum or damar gum, and thus it is a more efficient weighting agent. It is permitted for use in several countries (e.g., Canada, Sweden, Norway, and Brazil), but regulators in the United States and the European Community are waiting for the results of long-term safety studies. Given the structure of the molecule (and its obvious relationship to the recently approved Olestra), approval can be expected, but obviously no time frame can be predicted for such action.

Centipoise—An older unit of viscosity (1 centipoise = 1 mPa·sec [SI unit], the approximate viscosity of water at room temperature).

EMULSION STABILIZERS

Gum arabic is the most widely used stabilizer for beverage emulsions. As discussed in Chapter 6, a relatively concentrated solution has a low viscosity, facilitating the production process. A 30% solution in water (a typical concentration for making these emulsions) has a viscosity of less than 100 *centipoise* (about the viscosity of vegetable oil). The hydrophilic/lipophilic balance (HLB) of gum arabic has been reported as 8 and as 12. The discrepancy is not surprising (applying the HLB scale to a gum takes it somewhat outside its original design purpose), but either number indicates that gum arabic should be a modest to good emulsifier for O/W emulsions. While gum arabic lowers interfacial tension, its effectiveness as a stabilizer results mainly from the interfacial film it forms and the resultant high interfacial viscosity.

Gum tragacanth is also sometimes used in beverage emulsion systems. It is a medium-viscosity gum, but in combination with gum arabic, the viscosity of the overall solution is less than that of either of the gums alone. However, the two in combination form a mixed interfacial film, providing a more effective stabilizer than either gum alone.

A particular type of modified starch, which has been reacted with a derivatized succinic anhydride, is also suggested for use as a beverage emulsifier. The derivative consists of a lipophilic alkyl or alkenyl chain located at carbon 2 of the succinic acid. In the United States and several other countries, modified starch substituted with 2-(1-octenyl)-succinic anhydride (Fig. 7-2) has been approved for food use. The starch has at least two advantages over gum arabic as an emulsifier for beverage systems. First, as a modified corn starch, it is "cleaner" (i.e., less likely to have off-flavors or odors) than the average gum preparation. Second, because of the added lipophilic character resulting from the substituent molecule, less modified starch is needed for stabilization of the beverage emulsion.

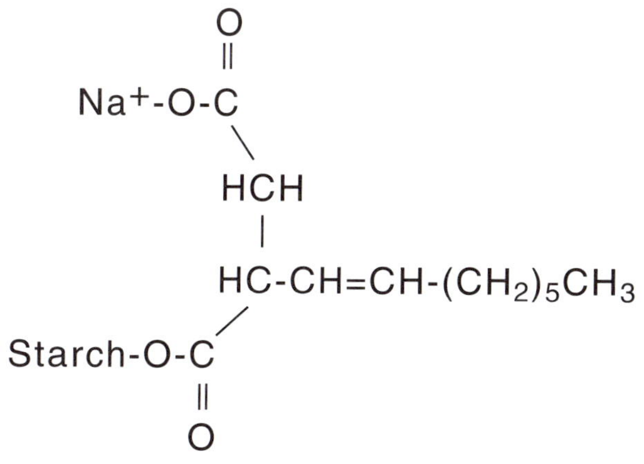

Fig. 7-2. Structure of 2-(1-octenyl) succinic acid, esterified to a hydroxyl group in starch to produce a modified starch for flavor emulsion stabilization.

EMULSION PREPARATION

Making a beverage emulsion follows the basic principles of emulsion production. The oil phase (flavors and weighting agents) is prepared and mixed with two to four times its volume of the aqueous phase. A mixer is used to produce a crude emulsion with droplet diameters of approximately 20 µm. The crude emulsion is then homogenized (in an apparatus similar to that used for homogenizing milk), reducing the average droplet diameter to 0.5–1 μm. This emulsion is packaged and shipped to the bottler.

Stability

Instability in emulsions is evidenced by three phenomena: creaming, flocculation, and coalescence. Beverage emulsion concentrates may exhibit all three, but in the finished, bottled beverage, only the first is likely to be noticed by the consumer after the drink has been stored for a period of time.

CREAMING

Creaming occurs when the dispersed phase of an emulsion is lighter than the continuous phase and the dispersion remains quiescent for a period of time. The collected, oil-rich cream layer is usually easily redispersed by gentle agitation of the total container contents. In soft drinks, creaming is often referred to as "ringing," because the flavor oil droplets collect around the neck of the bottle in a creamy ring.

The rate at which the particles rise is determined by Stokes's law:

$$v = 2gr^2(\rho_1 - \rho_2)/9\eta$$

in which v is the rate of creaming (or sedimentation, if the droplets are heavier than water), g is the gravitational constant, r is droplet radius, ρ_1 and ρ_2 are the densities of the oil and water phases, respectively, and η is the viscosity of the water phase. The water viscosity can be considered essentially constant, although it is slightly higher for a 10–12% sugar solution than for a diet drink in which a nonnutritive sweetener is used. Likewise, g is a constant factor. Thus, phase densities and droplet radius are the manipulable parameters.

The densities of 10 and 12% sugar solutions are 1.04 and 1.048 g/cm^3, respectively. The only notable difference is in a diet drink, where the density is about 1 g/cm^3. As mentioned earlier, citrus oils have a typical density of 0.85 g/cm^3. This density is increased by the addition of weighting agents, but the amount of these agents that can be used is regulated. Using the maximum level of weighting agent allowed by law raises the oil density to about 0.95–0.98, which means that the droplets still rise in the beverage. Another factor, however, is that the stabilizer (gum arabic or modified starch) is heavier than water. Thus, the interfacial film of the stabilizer used also increases droplet density. Because the exact film thickness is not generally known and because the ratio of oil to polysaccharide varies with droplet size (oil volume varies as the cube of the radius, while surface area varies as the square of the radius), a precise value for overall droplet density cannot be calculated. This is one reason that a gum stabilizer is preferable to a small-molecule emulsifier (such as a polysorbate).

Droplet radius is reduced as much as possible by the homogenization step. In studies in which some ringing was observed, comparisons of the particle size of the collected material with that of the droplets still in the main body of the drink showed that the majority of the creamed particles were 1–8 µm in diameter, while the suspended particles were nearly all less than 1 µm in diameter. A creaming velocity of 1 µm/24 hr is usually offset by Brownian motion (random thermal convection) in the drink, and no ringing is seen. In a practical sense and given ordinary values for oil droplet density, this means that droplets with a radius of less than about 0.5 µm will not produce a visible ring in the bottle neck.

FLOCCULATION

In a concentrated emulsion, where the dispersed phase may constitute 20–35% of the total volume, the droplets make contact. The hydrophilic layers interact loosely through van der Waal's and ionic forces, and a loose, low-density floc is formed. These aggregates, or clumps, act like large droplets and tend to rise more rapidly than individual drops. The sizes of individual drops remain unchanged; there is no coalescence of two drops into one. Flocculated emulsions, like creamed ones, are fairly readily redispersed by simple agitation, although flocculated emulsions may require a bit more vigor to accomplish this. The only practical concern caused by flocculation occurs at the bottling plant. The workers must be sure that each package (drum or carboy) of concentrated emulsion is thoroughly mixed before the emulsion is introduced into the bottling line; otherwise, the flavor level may be inconsistent in the bottles or cans of drink.

COALESCENCE

Gum (or modified starch) emulsion stabilizers form a film around the oil droplets. Such films develop a definite viscoelastic nature after a few days and can be observed macroscopically. A drop of oil is extruded into a gum solution and allowed to age for several hours. Some of the oil is then withdrawn with a syringe and a needle, and a crumpled envelope is seen. The phenomenon is much like that shown by α-tending emulsifiers, discussed in Chapter 3. Because of this interfacial film, true coalescence of oil droplets is seldom seen in flavor emulsions.

Microemulsions

The conditions necessary for spontaneous emulsification (i.e., the formation of microemulsions) are discussed in Chapter 1. Such emulsions would seem to be natural choices for the citrus-flavored beverages in which transparency is desired. A great deal of research has been done in this area, mainly in the laboratories of companies interested in producing such beverages, but little or nothing has yet appeared in the marketplace.

The main problem is to find two food-grade cosurfactants that produce the required low interfacial tension. Some that have been explored include acetyl monoglyceride and polyglycerol esters, but the usual cosurfactant has been something like a medium-chain-length alcohol (e.g., hexanol), which is not allowed in foods. Further fundamental work is likely to reveal a combination of surfactants that will produce acceptable food microemulsions, but none are commonly known at this time.

Troubleshooting

Emulsifiers are not used in beverage flavor emulsions because of the bitter flavors often noted. Emulsion stabilization depends primarily on gums.

Symptom	Causes	Changes to Make
"Ringing"	Large emulsion drops	Increase amount of gum. Improve homogenization methods.
	Low density of oil drops	Increase the amount of weighting agent (insofar as legally possible). Increase amount of gum.

Glossary

Acetal linkages—Bonds between sugar residues in polymers, linking the carbonyl group of one residue to a hydroxyl group on the other sugar.

Acetone insolubles—Specification of the amount of phospholipids in "gums," based on the fact that the other constituents normally present are soluble in acetone.

Acid value—Weight in milligrams of potassium hydroxide required to neutralize the titratable groups in 1 g of lipid. It characterizes a lipid by quantifying the proportion of titratable acidic groups.

Agglomerate—To remain in close proximity but, because of any of a variety of forces, not coalesce; e.g., individual particles in a suspension.

α-Monoglyceride—Monoglyceride in which the fatty acid is esterified to the 1 position of glycerol. Esterification at the 2 position results in a β-monoglyceride.

α-Tending emulsifier—Emulsifier that forms a solid film at the oil-water interface under proper conditions of temperature (low) and concentration (high).

Amorphous—Pertaining to the random arrangement of atoms or molecules with no discernible long-range order.

Amphiphilic—"Both loving." Pertains to molecules that possess both lipophilic ("fat-loving") and hydrophilic ("water-loving") regions.

Amylograph—An instrument used to study starch gelatinization. A slurry is heated from room temperature to 95°C at a set rate, held for a period of time, and then cooled at a set rate. The viscosity of the slurry is recorded as a function of time (hence, of temperature).

Amylopectin—Branched polyglucose chains in starch.

Amylose—Linear polyglucose chains in starch.

Bingham plastic—A suspension that does not flow when subjected to a shear stress lower than the yield value but does flow at any shear stress above the yield value.

Birefringence—Ability of crystalline materials to rotate polarized light.

Bloom—A dusty, whitish appearance of the surface of chocolate coatings, caused by transformation of the fat crystals.

Body—A qualitative measurement determined in sensory tests as the ability to "fill the mouth" with the characteristic being measured.

Brominated vegetable oil—Oil made by treating an unsaturated vegetable oil with molecular bromine. Addition of HBr across double bonds gives these oils a high density.

Brownian motion—The random, thermal movement of minute, solute particles observable under a microscope and caused by collision of solvent molecules with the particles.

Casein—The main protein component of milk, accounting for about 80% of the total proteins.

Centipoise—An older unit of viscosity (1 centipoise = 1 mPa·sec [SI unit], the approximate viscosity of water at room temperature).

Chorleywood bread process—A rapid bread-making process, developed at the Chorleywood Laboratories in England, in which dough is mixed in a high-intensity mixer for a short period of time.

Coalesce—To combine; usually refers to two liquid (e.g., oil) drops combining into one drop.

Colloid mill—Machine used to decrease the size of suspended particles or droplets.

Complexation—Combination of two different molecular species.

Continuous phase—The undispersed phase of an emulsion. In an oil-in-water emulsion, water is the continuous phase.

Critical micelle concentration (CMC)—Concentration of a surfactant in aqueous solution at which colligative properties cease to change with increase in concentration.

Crosslinked starch—Starch in which hydroxyl groups on adjacent chains are joined by a covalent linkage.

Crude gum—Material removed during the degumming phase of vegetable oil refining. Water is added to the crude oil, and the polar components (such as phospholipids) become hydrated and associated with the aqueous phase, which is then separated by centrifugation.

Crumb modulus—Synonym for elastic modulus.

Crystal inhibitors—Small molecules that interfere with the deposition of dissolved molecules on the growing face of a crystal, thus inhibiting growth of the crystal.

Crystalline—Pertaining to a state in which atoms or molecules are arranged in an ordered three-dimensional array. Long-range order is discerned by X-ray analysis.

Crystallites—Small regions of crystalline starch within a granule.

Cubic mesophase—Mesophase in which spheres of water are found in a cubical arrangement in a matrix of the surfactant.

Damar gum—A shrub exudate consisting mainly of triterpenes (containing 30 carbon atoms) but with numerous other related compounds also present.

Differential scanning calorimetry (DSC)—A method for measuring energy uptake as a sample is heated. When a phase change occurs (e.g., melting or freezing), the plot shows the temperature at which the change occurred and the amount of heat energy involved.

Differential thermal analysis—Method for measuring energy uptake as a sample is heated; similar to differential scanning calorimetry.

Diglyceride—Lipid with two fatty acids esterified to a glycerol molecule.

Discontinuous phase—The dispersed (internal) phase in an emulsion. In an oil-in-water emulsion, oil is the discontinuous phase.

Diterpenoid carboxylic acids—Tricyclic compounds containing 20 carbon atoms and with a carboxylic acid function attached.

Dough strengthener—Material added to bread dough to increase the ability of the gluten to retain gas during proofing and baking.

Egg albumin—Soluble protein found in egg white.

Elastic modulus—Relationship between stress (force) applied to a sample and the strain (deformation) in the sample; a more general rheological term than crumb modulus.

Electrical potential—The magnitude of electrical charge difference between two points.

Emulsifiers—Molecules that promote and/or stabilize emulsification, i.e., dispersion of one liquid in another (nonmiscible) liquid.

Essential oils—Aromatic and flavorous oils obtained from various plant sources and originally known as "essences," mainly in the perfume industry.

Ester gum—Gum made by esterifying the acids of purified wood resin with glycerol.

Fat mimetic—Fat substitute that mimics the properties of fat.

Feathering—Streaks of fat and precipitated protein that form in a liquid such as coffee when improperly stabilized whitener is added; caused by breakdown of the emulsion.

Flavor notes—Small molecules, usually volatile hydrocarbons, that interact with taste receptors and are experienced as flavors.

Flocculation—Collection of the internal phase of an emulsion or suspension; clumping. Flocculated materials are generally somewhat difficult to redisperse, unlike droplets in a creamed emulsion, which are easily dispersed by simple mixing.

Fluorescent probes—Small molecules that fluoresce in a nonpolar medium but not in a polar medium such as water. When these probes are mixed with proteins, for example, the appearance of fluorescence implies that a probe has bound to a hydrophobic region of the protein.

Free energy—The thermodynamic energy of a closed system. Absolute free energy is not easily measured, but the change in free energy when the system is changed (e.g., when oil is dispersed in water) is more easily established.

Freeze-thaw stability—The ability of a product to be frozen, thawed, and refrozen several times without noticeable changes in physical characteristics.

Gelatinization—Collapse (disruption) of molecular orders within the starch granule manifested by irreversible changes in properties such as granular swelling, native crystalline melting, loss of birefringence, and starch solubilization.

Gelatinization temperature—A narrow temperature range at which starch granules begin to swell, lose crystallinity, and viscosify the cooking medium. Starches from different sources have different characteristic gelatinization temperatures.

Glucopyranoside—Glucose in its usual molecular form of a six-membered ring.

Glycolipid—Diglyceride connected to a sugar moiety (usually galactose or galactosyl-galactose) at the 3 position and common in cereal and legume seeds.

Gum—Polysaccharide that markedly increases viscosity when dissolved or dispersed in water.

Helix—Three-dimensional arrangement of many biological polymers, including starch. It is analogous to a coil spring.

Hexagonal II mesophase—Mesophase in which cylinders of water are arranged hexagonally in a matrix of the surfactant. When cylinders of surfactant are present with water as the matrix, it is called a hexagonal I mesophase.

Hexose—Sugar containing six carbon atoms.

Hydrophilic—"Water loving." Pertains to the (polar) parts of molecules that readily dissolve or disperse in a polar medium such as water.

Hydrophilic/lipophilic balance (HLB)—Ratio of an emulsifier's hydrophilic and lipophilic tendencies.

Hydrophobic patches—Areas on the surface of a protein molecule, in contact with surrounding water, that are lipophilic in nature.

Hydrophobic—"Water hating." Pertains to the (nonpolar) parts of molecules that do not readily enter a polar medium such as water.

Interfaces—Boundaries between two phases. Various types of interfaces occur in foods: solid-liquid, gas-liquid, gas-solid, and liquid-liquid (two immiscible liquids).

Interfacial viscosity—Resistance to flow in the two dimensions of the interface.

Iodine value—A measurement of the number of double bonds in a fat or oil. A higher value means more double bonds.

Ionic strength—Measure of the ionic character of an aqueous solution of salts. Ionic strength plays a role in numerous physical phenomena such as conductance of electrical current, folding of protein molecules, and degree of repulsion of charged surfaces in water.

Isosorbide—Bicyclic structure formed from sorbitol, involving an ether linkage between hydroxyls at the 1 and 4 positions and hydroxyls at the 3 and 6 positions.

Lamellar mesophase—Mesophase characterized by bilayer leaflets of surfactant separated by layers of water.

Lipophilic—"Lipid loving." Pertains to the nonpolar parts of molecules that dissolve readily in a nonpolar medium such as vegetable oil. Generally synonymous with "hydrophobic."

Lipoprotein—A complex of a protein with a lipid (generally a phospholipid) found in many plants. The most common in food processing is from egg yolk.

Liposome—Hollow sphere formed by bilayers of surfactant, suspended in a water matrix, and containing water in the center.

Margarine—A substitute for butter, originally formulated to mimic butter as closely as possible, but using other fat sources in place of milk fat.

Mesophase—Opalescent or transparent liquid formed by a mixture of surfactant and water.

Metastable—Pertaining to a physical state that is not at the lowest possible free energy for the system. The activation energy for transition to the more stable state is high enough that the system remains in the current state for a significant length of time.

Microemulsions—Emulsions in which the diameters of the droplets in the dispersed phase are much smaller than the wavelength of visible light. The droplets do not scatter light; hence, the emulsion appears transparent.

Milk fat—The natural fat found in milk consisting of a mixture of glycerides.

Monoglyceride—Lipid with one fatty acid esterified to a glycerol molecule.

Monoterpenes—Aliphatic compounds, containing 10 carbon atoms, formed biologically by combination of two molecules of isoprene (2-methyl butadiene), with subsequent modification to give the various members of this group.

Oligosaccharide—Short polymer of sugar having three to eight sugar units.

Ovenspring—Increase in the volume of a loaf of bread during baking; i.e., the final loaf volume minus the volume of the dough at the end of the proof.

Overrun—Increase in volume of a whipped material resulting from the incorporation of air.

Pentose—Sugar containing five carbon atoms.

Phase diagram—A method of showing which mesophase structures are present at various water concentrations and temperatures.

Phase inversion—Conversion of the continuous phase of an emulsion to the discontinuous phase and vice versa.

Phospholipid—Diglyceride esterified at the 3 position to phosphoric acid, which in turn is often esterified to another group.

Photon—Basic unit of light waves.

Plant exudate—Droplets of hardened gum exuded by a plant to seal a break in the bark.

Plateau's border—The point at which three or four gas cells nearly touch in a foam.

Polymorphic behavior—Ability of a material to crystallize in more than one three-dimensional arrangement.

Polypeptide—A polymer consisting of amino acids connected by an amide bond, involving the carboxylic acid and α amino groups.

Polysaccharide—A carbohydrate containing several hundred, thousand, or hundred thousand sugar units (from the Greek *poly*, meaning "many").

Primary hydroxyl—Hydroxyl group on the terminal carbon atom of a compound. That carbon atom is connected to only one other carbon.

Response surface methodology—Method of experimental design to determine the optimum values for two or more variables in a product by using a limited number of experiments and making interpolations based on the experimental data.

Retrogradation—Recrystallization of gelatinized starch.

Saponification value—Weight in milligrams of the potassium hydroxide required to saponify 1 g of a lipid. It characterizes a lipid by quantifying the proportion of ester groups relative to the total molecular weight.

Secondary hydroxyl—Hydroxyl group on a carbon atom that is attached to two other carbons.

Sesquiterpenes—Similar to monoterpenes but containing 15 carbon atoms; formed by reaction of three isoprene units.

Shear thinning—Decrease in the viscosity of a suspension as shear rate increases.

Solid fat content—A measure of the amount of solid fat in a fat at various temperatures, determined by nuclear magnetic resonance.

Sorbitan—Cyclic structure formed by linking the hydroxyls at the 1 and 4 positions of sorbitol through an ether linkage.

Sorbitol—Sugar alcohol produced by reduction of glucose, consisting of a chain of six carbons with a hydroxyl group on each carbon.

Specific volume—In baking research, the weight of the cooled loaf divided by its volume.

Stabilized starch—Starch in which hydroxyl groups form ester or ether bonds with other small molecules.

Staling—Phenomenon that occurs in baked products during storage. Stale product has a firmer crumb structure than fresh product; the crumb has a dry, harsh texture; and the flavor impact is significantly reduced.

Starch granules—Naturally occurring, partially crystalline, discrete aggregates of amylose and amylopectin.

Sucrose acetate isobutyrate—The octaester of sucrose with acetic and isobutyric acids.

Supercooled—Pertaining to a solution that is cooled below the temperature at which crystals would ordinarily begin to form but do not because of the absence of nucleation or the presence of crystal inhibitors.

Surface excess—The concentration of a surfactant in the interfacial region compared with its concentration in the bulk phase in which it is dissolved.

Surface tension—The component of total free energy in a closed system caused by the presence of an interface.

Syneresis—Separation of a liquid from a gel; weeping.

Synergistic—Pertaining to a combination of two materials that displays more functionality than would be expected by simply summing the individual functionalities of the materials.

Thermodynamic activity—A factor accounting for the fact that the concentration-dependent properties of dissolved molecules often deviate from a strictly linear dependence. For surfactants below the critical micelle concentration, this deviation is usually negligible.

Triglyceride—Lipid with three fatty acids esterified to a glycerol molecule.

Unit cell—The smallest ordered unit of a crystal. It may contain only a few atoms (e.g., sodium chloride) or several molecules (e.g., a fat unit crystal).

Uronic acids—Derivatives of a sugar, in which the terminal -CH_2OH group is oxidized to a carboxylic acid. D-Glucuronic acid is derived from D-glucose, D-mannuronic acid from D-mannose, and D-galacturonic acid from D-galactose.

van der Waal's forces—Short-range attractive forces between molecules resulting from the dipole moment of atoms in the molecules.

Viscous isotropic phase—Another name for the cubic mesophase when the amount of water is sufficient to significantly raise the viscosity above that of the melted surfactant alone.

Visible light—Light visible to the human eye. The wavelength of visible light is approximately 400–700 nm.

Volume fraction—The fraction of total volume of a system represented by one of the discrete phases.

Waxy maize starch—Starch from genetically modified corn; contains almost no amylose.

Weighting agent—An oil-soluble material that has a density greater than that of water.

Wetting agents—Agents that promote spreading of a liquid on a solid surface.

Whey—The liquid left after casein has been precipitated from milk. In addition to protein, it contains lactose (milk sugar) and ash (inorganic salts).

Winterize—To cool salad oil until high-melting-point triglycerides form crystals. These crystals are removed so that the next time the oil is cooled, the cloudiness that results from crystallization will not occur.

X rays—A region of light waves with short wave length and high-energy photons.

Index

Abietic acid, 90
Acetone insolubles content, 35
Acetylated monoglyceride (AcMG), 28, 29, 62
Aeration agents, 61–63
α-Tending emulsifiers
  crystal form, 17
  interfacial film formed by, 11
  structures, 28
  uses
    aeration, 62
    cake production, 29
    nondairy whipped toppings, 73
Amylopectin, 48, 51, 57
Amylose, 48, 49, 51, 56, 57
Anionic emulsifiers, 33–34
Antistaling agents
  in bread, 55–56
  effects of, on starch
    complexation, 56–58
    gelatinization, 48–51
    retrogradation, 51–55
  monoglycerides, 20, 23, 26
Avrami equation, 52–53

Bingham plastic, 81, 82, 83
Bloom, 18
Bread staling, 51–56
Brownian motion, 12
Butter, 68–70

Cake batters, 61–63
Calcium stearoyl lactylate, 60
Capillary rise method, 5
Cellulose, 80, 81
Cetyltrimethylammonium bromide (CTAB), 2
Clouding 17, 84
Coalescence, 7, 93
Coffee whiteners, 74–75
Colloid mill, 83, 84
Cookies, 63
Creaming, 10, 92–93
Critical micelle concentration, 4, 18–19, 42
Crude gum, 34
Crystal inhibitors (modifiers), 17–18
  in ice cream, 74
  in salad dressings, 81
Crystal structures, of mono-, di-, and triglycerides, 15–17
Cubic mesophase, 19, 20, 21

Damar gum, 90
Decaglycerol dioleate, 38
Decaglycerol hexaoleate, 38
Decaglycerol monooleate, 38
Decaglycerol monostearate, 38
Diacetyl tartrate ester of monoglyceride (DATEM), 27, 28
  antistaling effects of, 60, 61
  as a dough strengthener, 59, 60
  effects of, on starch gelatinization, 50
Diacetyltartrate esters, 20
Differential scanning calorimeter, 49
Diglycerides, crystal structure of, 16
Dough strengtheners, 58–61
  structures, 27–29
  uses, 33–34
Drop weight, 5

Eggs, 82, 83
Electrical potential, 8, 9
Emulsions
  in beverages, 91–92
  flavor emulsions, 89–91
  formation, 6–9
    microemulsions, 14, 93–94
    spontaneous emulsions, 14, 93
  stabilization, 9–11, 91, 92–93
Essential oils, 89
Ester gum, 90
Ethoxylated monoglyceride (EMG)
  as a dough strengthener, 60
  hydrophilic/lipophilic balance, 38
  structure, 27, 28, 30

Fat
  effects of, on flavor, 86
  mimetics, 85
Flavor emulsions, 89–91
Flavor notes, 86
Flocculation, 9, 11, 93
Fluid isotropic mesophase, 20

Foams
  drainage, 11–12
  formation, 8–9
    nondairy whipped toppings, 72–73
    whipped cream, 71–72
  stabilization, 39, 43
Free energy. *See* Surface free energy *and* Interfacial tension

Glycerol monolaurate, 38
Glycerol monostearate (GMS)
  antistaling effects, 60, 61
  composition, 25, 26
  crystallization, 16
  effects of, on starch gelatinization, 50, 51
  hydrophilic/lipophilic balance, 38
  interfacial tension, 4
  lamellar mesophase formation, 19, 20
  phase diagram, 22
  uses
    coffee whiteners, 74
    dough strengthener, 60
Gums
  in beverages, 90, 91
  in dressings and sauces, 77–79, 81
  in reduced-fat products, 85

Hexaglycerol dioleate, 38
Hexagonal II mesophase, 21
Hydrogen bonding, 16, 17
Hydrophilic/lipophilic balance (HLB), 30, 36–39
  of gum arabic, 91
  of lecithin, 36
  of polyglycerol esters, 32
  of sucrose esters, 33

Ice cream, 73–74
Interfacial tension, 3–4, 7
Interfacial viscosity, 43
Ionic stabilization, 10
Ionic strength, 2

Lactic acid, 29, 33
Lactylated monoglyceride (LacMG)
  and aeration, 62

structure, 28
in whipped cream, 72
Lamellar mesophase, 19–20, 21
Lauryl sulfate, 2
Laurylsulfonic acid, 4
Lecithin, 2, 34–36
lamellar dispersion, 21
as a wetting agent, 13
Lipoproteins, 40
in mayonnaise, 84
in salad dressings, 82
Liposomes, 21

Margarine, 70–71
crystal inhibitors in, 17–18
Mayonnaise, 83–84
Mesophases, 18–21
phase diagrams, 21–22
Micellization, 18–19, 41
Microemulsions, 14, 93–94
Milk, 67–68
Monoglycerides, 2, 25–27
crystal structure, 15–16
derivatives, 27–30
hexagonal II mesophase formation, 21
phase diagrams, 22
polymorphic behavior, 16–17
uses
aeration, 61
antistaling effects, 20, 23, 26
ice cream, 74
shortenings and margarines, 18
Monostearin. See Glycerol monostearate

Nondairy whipped toppings, 72–73

Oriented wedge theory, 8
Oxystearin, 17, 81

Pendant drop, 5
pH, effects of, 38
on gluten protein, 58, 59
on viscosity, 79, 80
Phase diagrams, 21–22
Phase inversion, 68, 69
Phosphatidylcholine (PC), 35, 36
Phosphatidylethanolamine (PE), 35, 36
Phosphatidylinositol (PI), 35, 36
Plateau's border, 12
Polyglycerol esters, 17, 31–32, 81
Polyglycerol monostearate, 32
Polyoxyethylene sorbitan esters, 17
Polysaccharides, 77–80
Polysorbates, 21, 23
Polysorbate 60, 30, 31
hydrophilic/lipophilic balance, 38
uses
aeration, 62, 72
dough strengthener, 60

nondairy whipped toppings, 73
salad dressings, 81
Polysorbate 65, 30, 31
hydrophilic/lipophilic balance, 38
uses
coffee whiteners, 74
whipped cream, 72
Polysorbate 80, 30
hydrophilic/lipophilic balance, 38
uses
ice cream, 74
whipped cream, 72
Potassium oleate, 38
Propylene glycol monoester (PGME)
production, 29
structure, 28
use, 62
Propylene glycol monolaurate, 38
Propylene glycol monostearate, 11, 38, 73. See also Propylene glycol monoester
Proteins
as emulsifying agents, 39–43
as foaming agents, 39
in milk, 67–68

Reduced-fat dressings and sauces, 85–86
Regulations, 43–44
Ring tensiometer, 5–6

Salad dressings, 80–83
Shear, effects of, 6, 7, 72, 80
Shortening, 17–18
Sodium chloride, 9, 10
Sodium dodecyl sulfate (SDS), 33, 34, 41, 60
Sodium lauryl sulfate (SLS), 33, 34
hydrophilic/lipophilic balance, 38
uses
wetting agent, 13
whipping aid, 63
Sodium oleate, 38
Sodium stearate, 7, 8
Sodium stearoyl lactylate (SSL), 33, 34
effects of, on viscosity, 50, 51
hexagonal II mesophase formation, 21
hydrophilic/lipophilic balance, 38
uses
dough strengthener, 23, 58, 59, 60
coffee whiteners, 75
Sodium stearyl fumarate, 33, 34
Solid fat content, 69
Sorbitan monostearate, 30
as a crystal inhibitor, 17, 18
hydrophilic/lipophilic balance, 38
production, 25
Sorbitan tristearate, 30
as a crystal stabilizer, 18
hydrophilic/lipophilic balance, 38

Specific volume, 54
Spontaneous emulsification, 14, 93
Starch, 91
complexation with emulsifiers, 56–58
gelatinization, 48–51
modification, 79–80
retrogradation, 51–55
waxy maize starch, 82
Stearic acid
in production of food emulsifiers, 25, 29, 30, 33
structure, 2
Stearoyl lactylic acid, 62, 63
Steric hindrance, 10–11, 12
Succinate esters, 20
Succinyl monoglyceride (SMG)
as a dough strengthener, 60
structure, 27, 28
Sucrose acetate isobutyrate, 90
Sucrose esters, 32–33
and aeration, 62, 63
as crystal inhibitors, 17
Sucrose monoester, 38
Sucrose triester, 38
Surface excess, 4–5
Surface free energy, 3–4
Surface tension
in foams, 43
measurement, 5–6

Temperature, effects of
on emulsifier film formation, 62
on starch gelatinization, 49, 50, 51
on starch retrogradation, 54
on mesophase structure, 21–22
Thermodynamic activity, 4, 5
Triglycerides
crystal structure, 15–16
crystallization, 17
Triglycerol monostearate, 18, 38

van der Waal's forces, 8, 10, 20
Viscosity
and emulsion stability, 12
and gums, 77, 78
Viscous isotropic phase, 20
Volume fraction, 11

Water, effect of, on mesophase structure, 21–22
Weighting agents, 89, 92
Wetting agents, 13
Whipped cream, 71–72
Wilhelmy balance, 6
Winterization, 17, 81

Yield value, 82, 83

Zinc distearate, 7, 8
Zinc sulfate, 9